RAISING WATER BY FIRE
The Steam Letters

Judith Brooks

UK Book Publishing

RAISING WATER BY FIRE

First published in Great Britain as a softback original in 2019

Copyright © Judith Brooks

The moral right of this author has been asserted.

All rights reserved.

No part of this publication may be reproduced, stored in a retrieval system, or transmitted, in any form or by any means, without the prior permission in writing of the publisher, nor be otherwise circulated in any form of binding or cover other than that in which it is published and without a similar condition including this condition being imposed on the subsequent purchaser.

Typeset in Goudy Old Style

Design, typesetting and publishing by UK Book Publishing

www.ukbookpublishing.com

ISBN: 978-1-913179-16-8

Illustrations by Angelo Madrid
Cover: Whitbread Brewery Chiswell Street by George Garrard 1792
Courtesy of the British Library

'It was a real philosopher who could invent the fire engine, and first form the idea of producing so great an effect by a power in nature which had never before been thought of.'

ADAM SMITH L.L.D. AND F.R.S. 1776

Formerly Profeffor of Moral Philosophy in the University of Glasgow.

The Tyger

William Blake

1794

Tyger Tyger, burning bright,
In the forests of the night;
What immortal hand or eye,
Could frame thy fearful symmetry?

In what distant deeps or skies,
Burnt the fire of thine eyes?
On what wings dare he aspire?
What the hand, dare seize the fire?

And what shoulder, & what art,
Could twist the sinews of thy heart?
And when thy heart began to beat,
What dread hand? & what dread feet?

What the hammer? what the chain,
In what furnace was thy brain?
What the anvil? what dread grasp,
Dare its deadly terrors clasp!

When the stars threw down their spears
And water'd heaven with their tears:
Did he smile his work to see?
Did he who made the Lamb make thee?

Tyger Tyger burning bright,
In the forests of the night:
What immortal hand or eye,
Dare frame thy fearful symmetry?

Author's Note

Raising Water by Fire is not meant as an engineering text book though you cannot tell the story of steam without pistons, cylinders, valves and other whatnots making an appearance, those same tools that inventors cursed and loved through their fantastical dreams, sweat, struggle and death. While the dates, places, family details and key events follow history, the entries are my own, imagined as a chronological collection of diary entries, memoirs, or personal notes by the inventors themselves. Andrew Cosgrove, the blacksmith, and his family and acquaintances are fictional characters however the events they encounter are based on events of the time. The spelling used by different authors is only meant to give a flavour of the time and hopefully, does not interrupt meaning. My story begins with the horrors of a flooded mine and follows the invention of a pump remedy which changed the world.

PROLOGUE

You would have been amazed at the size of them once they were dug in. Their foundations were compacted down as if for a small house. Their supporting walls were double bricked, mortared and plumbed true as a scaffold for their great beams and cast-iron struts, braces, bolts and chains, all rough-hewn to modern eyes and shaped from forges burning like new suns through the night.

There were long heavy cylinders made by the best ironmongers in the world, cast in three pieces and joined by driven rivets, creating a long, breathing lung of steam holding its airtight piston and condenser as close as a winning hand of cards. Looming above these organs was a huge reciprocal beam of young-grained oak bound by iron straps stark against green hills like a reimagined siege machine.

These first steam engines were assembled on site by a long line of contractors delivering their wares and rubbing their hands in the morning cold, excited by the audacity before them and happy to see what may come of it. Perhaps the engineer's blood stirred as he climbed the ladders to the walkways and platforms and stood yards high above a mine's dark mouth or a canal's busy locks, his mind not on the green horizon but sensing the balance of the whole engine as a captain notes the lie of his ship. Soon he hoped to take the world by its collar and shake it to see what might fall from its secret pockets. He already knew the four elements would rise as he called air, earth, fire and water to confound his critics. But he did not know the Earth's axis itself would alter in a future time when men could measure such motion, causing some to suppose the vast ice shelf of Greenland had finally shifted and begun its slow, inevitable slide into the Atlantic Ocean.

When the giant *Invention for Raising Water by Fire* began its primal rush of motion independent of all previous forces known to them, the men watching laughed incredulous at its power while a world well beyond their imagining began to form its fabulous, arrogant self. This is how our age began, full of awe and coal dust: the swift and ruthless Anthropocene.

But those laughing men, innocent of such knowledge, knew only that a slow-burning fuse had changed their world, and those with young blood and high aspirations were eager to find that all things were possible.

This is the story of the men who led a revolution, and in particular Denis Papin, Thomas Newcomen and James Watt and of course the Lunar Society, whose genius stirred the making of the modern world.

Through the moving beams of a steam engine, if you can imagine a distant field, you might glimpse a family driven to track by horse and dray the rutted lanes that passed for roads, on their way to the nearest town where 'hands' were wanted in the new manufactories. They carried the skills of centuries of lace-makers, tailors, potters, millers and cobblers sealed in their memories. They had left the gardens they tended and cottages they worked in and their numbers were legion. They became the working poor and their labour would make the modern world.

*

REALISATIONS

PART 1

Steam Team 1

Robert Boyle 1627 - 1691

Christiaan Huygens 1629 - 1695

Denis Papin 1647 - c.1713

Thomas Savery 1650 - 1715

Thomas Newcomen 1664 - 1729

1

'The Temple Wonder'

Steam power was no stranger to anyone.
It hid in plain sight for thousands of years
waiting patiently to be unleashed.

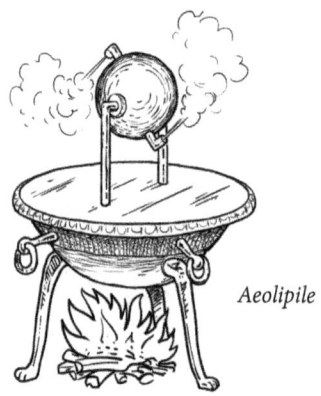

Aeolipile

Hero of Alexandria
Egypt, 100 CE

I do not know what to say of this marvel. It could be novelty, or a magical device or a Temple Wonder. I cannot watch it whirl of its own device without joy leaping through my veins. Here it is, my aeolipile, the Ball of Aeolus, from Aeolus, the God of Air and Wind. It is a sphere of heated water, held on two uprights so it can spin. The force of steam emitted from its two opposite tubes compels it to spin. I call it a steam engine and any man can make one.

*

2

'We wiped their faces clean of the flood which hadst choked every man'

Steam needed a particular climate to prosper. It needed rational curiosity and finally, a millennium and a half after the Temple Wonder, rational curiosity arrived under the banner of Enlightenment. It also needed a pressing necessity. Elias Newcomen was witness to the necessity. He is 30 years old, educated as a Dissenter and a leading member of the Baptist Brethren. He is an ironmonger by trade with a business in Dartmouth and is expecting his first child who he will call Thomas.

Elias Newcomen
Dartmouth, 1663

It is late and the Winde is high and howls about and it could be Satan himself walking the Commons throwing his minions hither and thither and all of them in a fine humour. I wouldst no man read these whiddes for I be drench'd in such trembling my thoughts betray me. It was noon when I heard the call and took my horse forthwith and thought all wouldst be well for many came as one and yet we were too late and saw that the flood hadst swallow'd them and the earth hadst swallow'd them and they were taken from us and from their families and from God's Earth. But the Tinners wouldst not suffer this and took their picks and found each man lost. To look upon them our eyes were scald'd, for the foul air of hell blew from their faces so their own mothers wouldst pause to name them. Who will know the motion of their souls and whither they are bound? We wiped their faces clean of the flood which hadst choked every man and there lying broken were the brothers Caleb and Eli Parker who hadst the day before come to Dartmouth to buy their tools from me and told me the Proprietor hadst dug a deep shaft and if their arms proved true they wouldst be rich men, God willing. The other Tinners were not all known to me but I wept just the same and not one man of us could stop until their poor bodies were wrapped in proper shrouds as befitting Christian men. These whiddes fly forth from my hand to bemoan my shrivelled soul which flaps about in such a winde tonight with Admonishments to the Lord that He shouldst drown such men bending their backs for their families. What are we to do when waterwheels fail and foulness rises from the depths? If I speak truly, there is no answer and it has made my Faith in the Rightness of Things tipped up. The unseemly filth of the world is spilt willy-nilly from the adits, pouring down the gullies passed the waterwheels and out onto the land. I declare as I write this, which I will fold away directly, I will not let a son of myne bend his back in any Mine, neither tin copper or coal. The Newcomens will be men who walk on the face of the Earth and they will make their way in the light of day and if we starve so be it, and if we prosper, so be it!

*

3

Nullius in verba
'take nobody's word for it'

Over 200 miles to the north, Robert Boyle, the delicate, sweet-faced 14th child of the Earl of Cork, has inherited a regular income and time enough to study his favourite delights. Working from Oxford University he has recently published a landmark book on chemistry. He is a founding fellow of the Royal Society. He is 36 years old and is a heady mixture of Anglican piety and sharp intelligence.

Robert Boyle
Pall Mall Street
London, 1663

I am come down especially from Oxford, thankful that the roads were dry and I lodge with my sister Katherine who declares my Visit timely as she was fast forgetting she has a Philosopher brother. She threaten'd to cover me with Kisses if I remain'd a Stranger. As a man with a surfeit of sisters and long-tamed into their Plaything, I withdrew in happy horror. I retired to make notes for there is a singular Purpose to my visit. I have not open'd the window for fear of bad Air but I am most settled in the afternoon warmth which flows from the great ball of the Sun, through Space and Time and the weight of such Air, to spill on my desk.

*

I happily write that we are greatly Honoured and I have come specially to witness such. We are no longer the *virtuosi* [1] who meet unheralded in the drafty halls of Gresham College, but will be known as Fellows of the Royal Society, the first such society named in this kingdom, and affirm'd by the King's own hand in his Second most prestigious Charter which names London and affirms the singular importance of Natural Philisophy. Tho' I do not apprehend the King will relieve us of the drafts which chill our knees, His Patronage is to our Greater Glory!

Katherine, being so enamoured of the Natural World, has decided we should converse in English as it is the Native Tongue, but my companions would be aggriev'd to learn Latin is banish'd at Pall Mall Street. It is to great consequence we have sworn *Nullius in verba*, vowing to take nobody's word for it. By this motto, we will have no Master but Facts as they are shewn True by experiment. Our purpose is to hold the Common Good and seek remedies for the toils of this frail life and, tho' we be not Commanders to defeat great Armies, we would be heroick of Mind. I would agree with Aristotle: that the whole is not only other than the sum of its parts, but greater, and truly, we should expect great Victories when diligent Minds join to learn the secrets of Creation.

*

April, being the month of Resurrection and new life, has added a sunny Zest to our Promotion. Katherine has bought into my room an Invasion of Sweetpeas and Daisys which she says have floated into her vegetable garden and give her great Pleasure and I should behold them to make me attuned to Beauty so I may look upon ladies with more Satisfaction! I gave her a flourishing bow and thank'd her most kindly and said that flowers under a roof were exceptionally Radical but might be just what I needed. After she had left I put them in a corner for I had already begun to sneeze and drip and assum'd there could be no better Evidence of my unsuitability for Matrimony and thought myself a Coward for my bow but a Genius for my dripping!

*

Today I woke early to a rude Concoction which reached down Pall Mall. Roosters and barking dogs, carts and the rasping throats of horsemen making their way threw the lanes assailed my ears. What has happened to the Peace of the Sabbath I thought to make Inquiry - but stretch'd in fine bed linens. I will confess I miss my hard Oxford cot not one jot. I am later to St Martins to worship and listen for God's Wisdom above the hum of earthly Bodies.

*

I return'd most amazed. I found every Tom, Dick and Harry flaunting their Faith among a press of periwigs. Rouged Ladies sent me longing for a private space where the Air was merely temper'd by the fresh Stink of the river, and happily, mother Thames had seaweed drying in the Sun and seabirds circling high enough to make me breathe again.

London is as full as I have ever seen, I told Katherine. Surely Vicar Hardy is rejoicing for his flock swells in weight of persons and prestige? We are Blessed that we have such a city about us and such Promises as the King has made, I continued.

Katherine put aside her book. I hope that it falls so, she said. This city will be forever taken with the virtue of its Parliament, upon which I take great store. I pray the King has resolved to Honour it so.

So saying her face fell into doubtful shadow but I did not take up argument for we have bothe a loathing of fitful Rebellion.

Do not fear for Parliament, I replied. It has made a name for itself and no king will forget it. (2) At that she gave a smile.

For all that, sweet Hope abounds, and London hums, no, I said, it shouts, as does Oxford, with the spirit of peace now we have our Restoration.

We are certainly restored, Katherine replied, you will see how ramshackle dwellings have sprout'd like mushrooms over our fields and ruin good pasture.

My sister has a great Affection for London. She sought refuge fleeing from cruel Siege by the rebels of Athlone, and is now in Pall Mall Street where she is greatly favoured among literary circles. She jests that every man and woman seemed a child come with bright faces to make their Fortune and report they do not bend their heads so much but take their joy in ale-houses and their Ceremony in the Guilds. I do fear to count the bastard Children who may be upon us and how we will feed them but Katherine calls me Jeremiah, old before my time. We laugh to hear our years rattle round our heads and swear we are not yet ready to give way to Apprentices dancing down our streets.

*

I am of a mind to be back at Oxford for there is much to be done. Katherine has remind'd me I am not the only Philosopher who wakes with a plume in his hand. She has received a post from her star-gazing farmer Richard Towneley, that his colleague Henry Power plans to publish at last.

His title will be *Experimental Philosophy* she said, holding his letter towards me. It will prove beyond all gainsay the relationship between the pressure and volume of gas, apparently, among many other things, she added.

A plain enough title, I said, thinking it could be more enticing to his fellow Philosophers. And, rather late, I added since the world knows I have publish'd and proved their Hypothesis.

Katherine placed Towneley's letter among her papers.

Henry Power is no barnacle, she replied, and his title is Plain but entirely to the Point, for what is the Society to be about if it is not to shew the world by the Experiments of many Philosophers the truth of its Equations?

I did agree, for my sister is, by self-appointment, my special Weather-Cock to shew which way my Winde should blow.

If you recall, she continued, it is two years since Towneley braved the roads to confer with us in London. He was in Awe of your publication *New Experiments Physico-Mechanicall, Touching the Spring of the Air, and Its Effects*, in which you used an air pump to create a Vacuum.

I did recall and even tho' I had one foot out the door I am prompted back to my desk to write this account for Towneley has about him the Virtue of Bold Notions most worthy of Consideration.

He chose to visit in Winter's Darkness, but arrived in a gleen of sunshine, his cheeks rosy and boasting the warmth of his mufflements.(3) He declared, it 'rained pikels with the tines downwards'(4) for nigh on one hundred miles and swore he was a lucky man to survive being drown'd on the open road. Towneley is much my own age but I remember he carri'd such a robust Wit he caused my Irish spirit to stir and I feared I looked a towne-fed Spaniel to his eyes.

I recall he was welcom'd with a fine rost by Katherine to whom he shewed a great Civility, firstly as my sister Lady Ranelagh, and then as my Mentor and Colleague. He announc'd himself in awe of her Reputation and eager to learn from her any medical Remedies she could recommend and very shortly Lady Ranelagh had seated him at her work desk and was in happy Discourse about all Manner of Things. From time to time laughter burst from their nattering which could have displeasur'd me had I been so inclined, but Towneley was a stout enough fellow and Katherine's laugh a rare delight.

Robert, she called, Mr Towneley is of the opinion that measurement is everything, he even plans to measure the rain!

At this Towneley turned to me and said I might think him mad but he planned a systematic recording of Rainfall since he was convinced that Lancashire was the rainiest County in all our Isles. (5)

I will join you this instant if you can tell me about your adventures with air pressure, I replied.

Indeed Sir, he said, blushing somewhat at his previous boldness, I am at your service.

He spread his notes on my desk and described his experiments with a Barometer on Pendle Hill in the company of Henry Power. Katherine had warn'd me to not to joke that they did well to preserve their lives against the ghosts of witches on those wild hills (6) lest our guest take

Fright - but Towneley was already describing his experiments his face flushed with joyous Amination.

He showed me his markings where he had shewn a relationship betwixt the density of Air and its Pressure. And further, the relationship betwixt the Pressure and Volume of gas when gas is contained, as he described, in a closed system. He started to laugh when he saw my expression and declared, it is so, it is so!

I laughed too and I felt my blood warm as if it might be summer in the streets after all and not untidy snow that threaten'd to dim our light. I asked Katherine if she could order us more candles and a good wine which, of course she did and joined us to hear the story of Towneley's adventures on Pendle Hill and we knew, as we listen'd, that in Towneley's muddy notes was a Great Advance. Katherine's eyes glowed with the joy of it and we gabbled like excited children over a new Adventure and it was a very late hour before we came to our senses and acted like good Christian hosts.

Towneley asked if I might give some thought to expanding on his experiments and so in my second paper, newly publish'd, I made publick his hypotheses and advanc'd it further where I explained the function of Gas Law, demonstrat'd its Truth and published its Equation. The Equation shows that, as Volume increases, the Pressure of the gas decreases, but in proportion. Similarly, as Volume decreases, the Pressure of the gas increases. Being so demonstrat'd and publish'd it is now known as Natural Law, true into God's Infinity. As to the significance of an air pressure pump and Vacuum we are yet to fully discover. I will put my mind to it and encourage other Fellows. Its powers of Suction will surely lead to further wonders for in one discovery is the seed of another.

*

4

'They came to me carrying honey cakes'

Elias Newcomen
Dartmouth, 1677

I hath spent this day in Disputation with my daughters where I was at the mercy of their Righteousness which flew around the house dancing into every corner with Bells and Ribbons blessing itself. They came to me carrying honey cakes warm from the hearth and sang me a sweet hymn so I might be soften'd as on a fine summer night I could let the stars distract me from bandits on the Exeter Road. Hadst they been older and not singing like angels I might hadst scold'd them with my Rod for they answer'd me so directly I almost fell off my chair. They came to plead for their brother Thomas not to be sent away to the Blacksmith but to be kept for Grammar school for he was, in their eyes, the most sensible and clever brother in all of Dartmouth. When I told them that might be but Newcomens were not scholars but makers of money by trade they fear'd not in reply. They said, as from Proverbs 1:8-9, that as my children they were a Garland to grace my head and a Chain to adorn my neck and begg'd me to favour their appeal. Being so amaz'd at their learn'd reply I sat dumbfounded until I gather'd my wits and call'd them close and said they could write to Thomas and visit him and meanwhile sing for him every night and I wouldst see they were free to do so. Their mother came and took them before my Righteousness got the better of me and I scold'd them to honour their father and mother and I signed the papers for Thomas and sent him off to Exeter the next day on a trader's wagon with enough food to last for three days.

*

5

'the future may arrive dressed in unfamiliar clothes'

Robert Boyle
Pall Mall Street
London, 1679

I will remark upon yesterday for I stood with Hooke well powder'd and puff'd at Gresham College for a most Remarkable occasion. I was to introduce Dr. Papin where he was, at our recommendation, about to demonstrate his *Steam Digester*. Fellows are apt to stand in rowdy Anticipation until they are called to our weekly business and were making a fine hubbub. I told Papin we must wait until we were certain of our audience.

It is a fine thing to have a King most interest'd in the quirks and queries of our Experiments but it plays Havock upon a steady Mind that he may, on a whim, demonstrate his Fascination with a visit *impromptu* and click his tongue at our *tardy* proceedings. He is a man most easy and pleasant in trifles but a Stuart nonetheless. Poor Hooke, as Curator of Experiments, plies Charles with lists of our Endeavours so his interest may be well-sated in all Manner of Things from the mysteries of Gravity to the growth of Plants but is never privy to the timing of his visits.

I waited till Hooke signalled there was no sign of palace horses and waved my cane in a most strident manner which made the Fellows laugh and posit if I ever had claim to rapier Wit I would never have claim to rapier Skill. Having caught their attention, I described Dr Papin's work and the iron-cased orb before us as an *'engine for softening bone using steam pressure.'*

There was much clearing of throats as Papin stepped forward for many thought him Hooke's secretary and were surpris'd to have him before them. There were candles aplenty in the room but the shadows, even to my eyes, made Papin a gaunt Frenchman with a nose too long and a wig too high. I could see it was a daunting Task to stand before men whose Language was not his own and whose eyes were ready to find Fault. I had introduc'd him as a Great Mind sheltering from Catholic Tyranny and a most worthy Speaker, but many Fellows failed to hide their smiles as his dark, round, bubbling grub was revealed. There was a snake like Hissing and the air had a smell of Burning and some fear'd they could feel the Explosion that should be upon them at any moment. But the Fellows stood their ground and a clear Understanding turned to Admiration when the Force of Steam was shewn.

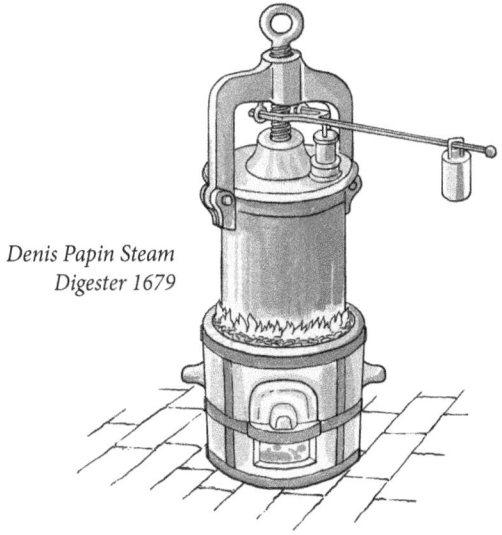

Denis Papin Steam Digester 1679

Despite his tumbling words, Papin's experiment could not be denied and I said as much to the Assembly. He shewed that steam alone trapp'd inside his pot raised the temperature 15 percentage points higher than the boiling Point of Water. The rate of cooking was remarkable and was receiv'd with general applause from the Fellows. Papin was silent on further uses for such power and I fear any greater use may prove to be a will-o'-the-wisp always receding from a philosopher's grasp.

*

I have been so bold as to make out a list for Experimentation as a guide to our studies. I have been warn'd by colleagues that I should do no such thing, for even among our friends many still laugh at what they consider our Absurdities. In reply, I would say that the Future may arrive dress'd in unfamiliar Clothes and knock unheard on many Doors, but it is a Philosopher's duty to carry Mankind through such Fears as Ignorance and Superstition may produce. Perhaps we should ask poets to write an Ode to Man's Curiosity for we might not always find what we seek, but are often led to unexpected Gems as happy Compensation!

My list is as follows (1)

The Prolongation of Life.
The Recovery of Youth, or at least some of the Marks of it, as new Teeth, new Hair colour'd as in youth.
The Art of Flying.
The Art of Continuing long under water, and exercising functions freely there.
The Cure of Wounds at a Distance.
The Cure of Diseases at a distance or at least by Transplantation.
The Attaining Gigantick Dimensions.
The Emulating of Fish without Engines by Custome and Education only.
The Acceleration of the Production of things out of Seed.
The Transmutation of Metalls.
The makeing of Glass Malleable.
The Transmutation of Species in Mineralls, Animals, and Vegetables.
The Liquid Alkaest and Other dissolving Menstruums.
The making of Parabolicall and Hyperbolicall Glasses.
The making Armor light and extremely hard.
The practicable and certain way of finding Longitudes.
The use of Pendulums at Sea and in Journeys, and the Application of it to watches.

Potent Druggs to alter or Exalt Imagination, Waking, Memory, and other functions, and appease pain, procure innocent sleep, harmless dreams, etc.

A Ship to saile with All Winds, and A Ship not to be Sunk.

Freedom from Necessity of much Sleeping exemplify'd by the Operations of Tea and what happens in Mad-Men.

Pleasing Dreams and physicall Exercises exemplify'd by the Egyptian. Electuary and by the Fungus mentioned by the French Author.

Great Strength and Agility of Body exemplify'd by that of Frantick Epileptick and Hystericall persons.

A perpetuall Light.

Varnishes perfumable by Rubbing.

*

6

'There is a smell about'

Christiaan Huygens is 61 years old and clinically depressed. Educated at the University of Leiden he is a brilliant mathematician, astronomer and inventor. When he was 26, while studying the rings of Saturn, he discovered its Saturni Luna, later named Titan. A year later he invented the pendulum clock which was not superseded as a timekeeper for 300 years. He is a Fellow of the Royal Society, London.

Christiaan Huygens
The Hague, 1690

There is a smell about which offends me, sour but pungent with its dankness. The sun and winde are my relief for otherwise it follows me and I recoil, thinking it is from the bad Humours that infect my Soul and lead me to dwell in Melancholy as if that dark Land is my rightful home on Earth. Lysbet scolds me about my complaints. She was my father's maid and has appointed herself my Housekeeper for she says no one else will tolerate me and she might as well serve the mad Son after the mad Father. She has decided the smell about is real enough and accuses my Feet of the cause and she does not need one of my Papers to prove her right. There is a bucket at my feet and a rough towel and she has vowed that even though she is a Christian woman and I a great Genius, I can wash my own feet.

I curse her in return for she makes her mirth at my expense and I cannot help but laugh with her. I have been two years in my father's house Hotwijck, named so because he liked to laugh with words; Hot being Court and wijck being Avoid. It is a lonely Inheritance where I am wijck the whole world. If it was not for the Post I would be entirely marooned in my father's summer Retreat with Lysbet and memories of childhood as ghosts in my Miasma.

It is a weight upon my heart that King Louis, once so enlightened, has turned against my faith and I am as welcome in Paris as a cadaver carrying the Plague. A King may take it upon Himself to be many things, but he should be, above all else, Reliable! Other lesser souls are so. This morning's Post brought news so welcome it broke me almost to weep.

I was delivered bound papers, sealed boldly Dr. D. Papin, Leipsig, Saxony. I have not seen my fellow Huguenot for many years but while the world about is changing I read that Papin has not. He has survived his exile teaching Mathematics at Marburg and has published from Leipsig, *De novis quibusdam machinis.* declaring this 'A New Method of Obtaining Very Great Moving Powers at a Small Cost.'

Without a moment of surprise, I read he has designed a Pump to raise water to a canal between Kassel and Karlshaven. He says he will do so by using the differences in heated pressures of Steam. He writes fondly to me that he has added his own Innovation to the work we did years ago. He says that water boiled under pressure is such a powerful Beast it will do instead of the Gunpowder we thought to use. I recall he took heartily to my suggestion that we should create a vacuum within a cylinder under a piston. He writes Steam could be used to Lift and Pull and Push all manner of things. I sat transfixed and ran my hand across his words. I had not thought of the gunpowder engine for many years but here today is Papin, tapping on my mind like a persistent goat desiring his breakfast.

This stirred my hunger and I washed my feet and called for coffee and cheese and settled to read his latest thinking for he was no ghost to me, this boy from Chitenay. He came as a protégé of Madame Colbert when Paris seemed the centre of the world. I was then Director and had about me a fine Academy of Science when France was a more Civil place. He was recently graduated with a medical degree but insisted he would be better put to making things than curing them. Gottfried Leibniz arrived shortly after, his Latin impeccable and his manners charming. He was another eager to escape his family's dictates, preferring the uncertain seas of Natural Philosophy to what he called the *capti per antrum* (1) of Diplomacy. I recall Leibniz as quick to learn and even quicker to challenge what he could not swallow. They both had quicksilver minds and holy passions to keep them company.

Leibniz offered me infinitesimal calculus for my perusal and I returned it unconvinced of its use! Our Patron, Minister Colbert, rejoicing in the Academy's future where the pupils proved the teacher wrong, slapped me on the back and declared he should make me a Medal for I was nurturing Cubs who would make France the leader of Europe. As if Memory can rise the Dead I see Colbert before me, waving his arms and declaring he wanted an entirely new form of power for France and I should work on the power of water converted by fire into steam. And now Papin writes from the wilds of Kessel, supported by Leibniz and his understanding of motive power. It is no surprise that they rub so well together.

Although he favours steam pressure he adds notes for what we called atmospheric pressure. He posits that steam condensed could create the vacuum we sort to force a Piston down a single Cylinder. He writes this is further to the work I suggested on Vacuum by the Late von Guericke of Magdeburg and the further work of his old friend, Robert Boyle. Without Von Guericke's *Experimenta Nova* of '72 (2) we would not know the nature of space and the possibility of the void. I was not a young man but still my heart raced at possibilities all fantastical like the Souls of the Unborn singing before their very Conception! For is it not true that new theories arrive as naked babes at the Mercy of the Established order but when they have been Proven they appear as being always naturally so?

The possibility of a Void opened a wider door and now Papin walks through that door and seeks to open many others. He deserves words of encouragement but I must first play Professor and read his detail. It will be of interest to him that I am still able to travel and was last summer in Cambridge at the invitation of Newton who is now a Senior Fellow at Trinity. Papin will have heard of Newton's extraordinary *Philosophiae Naturalis Principia Mathematica*,(3) for who has not, but he may not have heard how his erstwhile colleague Robert Hooke takes upon himself to accuse Newton of plagiarism regarding the theory of inverse squares. Hooke has little support. To have glimpsed an idea is only a milk slop in the making of a cake, and Hooke has never put forward any level of proof.

I will tell Papin I heard not one Philosopher in London, Cambridge or elsewhere discuss the potential of Steam power. Nothing had been done to advance his Steam Pressure Digester of '79 neither in Principle nor in Practice. Far from being discouraged he should find heart in this, for his persistence has made him a leader in this field and will, by and by, lead on to even greater discoveries.

As for Newton and I, we had a meeting of minds unblemished by the Noise of the World. We toured the quiet of his garden and took rest in the shade to discuss our mutual work on crystallised Calcium Carbonate hoping that the properties of this Crystal could reveal the nature of Light. Newton is still of the opinion that light is made of tiny particles whereas I posit that light is made from waves but we are of a mind to trust Experiment rather than Feuds to settle our theories.

It will be of interest also to Papin that he spoke highly of Robert Boyle, a man devoted to advancing knowledge but of poor health since he suffered a sudden paralysis. Newton is not the restless man I thought to meet and works happily in his *laboratorium*(4) and desires nothing more than to tinker like any Hatter with his mercury, dyes, and other extracts I could only guess at.

The day is drawing fast but I will walk for a time along the canal. I feel I could walk and walk. Lysbet will be amazed but I do not wish to be a prisoner of maudlin tears if I dwell on past times. The company of great Minds and great Visions should be drawn together for the

Universal Good, not spread hither and thither because of the Faith they hold, at the mercy of any Winde which blows.

I recall von Guericke's words where he argues God cannot be contained in any Location nor in any Vacuum nor in any Space, for He Himself is, of His Nature, Location and Vacuum. That may be, but I would have the Lord closer to home so He could sit by my fire and explain to me many wonders, the least of which is why I am on this Earth at His pleasure. Sometimes I fear He has told Lysbet and she is about to tell me His answer.

*

7

'a steam engine to raise water'

Robert Boyle
Pall Mall Street
London, 1690

 I am sickly most days and fear that paralysis, like a hungry shade, waited to pounce again on my limbs, so today's post was a Tonic for my spirit. I hold Dr. Papin's *De novis quibusdam machinis*, which he has sent to Pall Mall Street. It is a design for a steam engine to raise water to a canal between Kassel and Karlshaven by Differences in heated Pressures.

 It gives me great pleasure to know that steam is a hare Papin still pursues, despite the considerable Huffing and Puffing he made against the Society. He can be as changeable as a wind vane in a Tempest, yet that is of no real account. It is my apprehension that the duty of a man with a Great Mind is to advance his own genius and be not deterr'd by Dangers or discourag'd by Difficulties.[1] I would Papin be such a man.

 I write daily but not for long. My mornings become a Meditation which drift into a quiet sleep where I wake and care not what time is lost. In a pleasing way, the Past has crept back to me and presents itself disguis'd as Yesterday. Names I have not thought of for years may be the colleagues who softly knocked on my door.

 Robert Hooke has been my company this afternoon. We discussed Papin's Paper and spoke of our discoveries when all about us wad Theatre, Dash and Glory. Hooke's memory is far in advance of mine and he took pleasure recounting all our victories no doubt instruct'd by my sister to keep me in a cheerful disposition. And so I am, for *De novis quibusdam machinis* sits on my lap and I have a mind to pat it as I would a dog that might bark at me.

I wonder at its application and find I am more interested in his aside which speaks of an atmospheric pressure obtained by condensing steam. Papin suggests it is a force to be applied very advantageously and here he builds on principles more familiar. And to that purpose Hooke had me laughing like a schoolboy when he recall'd our astonishment many years ago when we heard of the German von Guericke's experiments with vacuum.

A space in which there is no Matter enthralled us beyond all imagining, I recalled.

It impressed us even more that von Guericke demonstrat'd it so boldly, he replied.

Yes. Was it not sixteen horses that failed to pull his hemispheres apart and break the Vacuum inside?

So it was, so it was, Hooke assured me. He forged a pair of copper hemispheres, sealed them with grease and pumped the air out. His men urged a team of horses, eight aside, but could not break the pressure of Air, he recalled.

The pressure of Air, and now the pressure of Steam, I mused.

We fell silent, or so it seemed. Perhaps Hooke thought it time for me to rest and retir'd quietly for I did not hear him go. I closed my eyes and judge'd most strangely I was in my workshop with Hooke. We were seeking the properties of air with Papin who smiles towards us. Here, he says, I have introduced Improvements to Hooke's air-pump by making it with double barrels and replacing the usual turn-cock with two valves, do you not think it an ingenious trick? I cannot recall my reply for there is a great Noise and Bustle from the kitchens and I have smooth'd the covers. Katherine will be up soon, anxious as to my mood and I will be delight'd to tell her it is a most fortunate one.

I plan to reply to Papin this very night. I will seek to give him support even though he is residing in Hesse a long way from the auspices of the Society. He was readily accepted when I nominat'd him for membership of the Society in '80, but a whole decade of work and struggle has passed by since.

His pen is often caustic for he writes that he carries the tag of Émigré, honourable in the eyes of God, but he does not hold his colleagues as Tolerant as the Almighty and he may have cause to think so.

I have held Papin's treatise in my hands for a great long while that day, musing on how my land lies fallow and I cannot look to leave the kind of hermetic Legacy I had planned. My energy is entirely used by waking, eating and shitting. If I did not have Katherine's happy intellect as my companion I would have long passed into Death's dark Kingdom. But today it would be a short Hypothesis to say the state of my Health is in equal inverse proportion to the Joy I received by the arrival of Papin's declaration of his Triumph over steam. I will caution him to proceed gradually for I would not have him fail for want of Patience, I fear for him and would caution him that he would not be the first man to discover there is a distance between a Truth that is Glimpsed and a Truth that is Demonstrat'd.[2]

If he were here I would have him listen to the bell ringers at St Sepulchre who have astonish all of London downwind of their marvellous peals. They call it Change ringing and it is a most Devious Delight in mathematical permutations. It could be a thousand bells it is so grand and they ring joy to every Soul.

I have in mind to title my notes The Christian Virtuoso, where I affirm that the wonder of Knowledge will assist man to love God. It has surely been so for me, I thought because I chose it to be so, but now I see God has planned it all along for the Natural World stands Testimony to the Genius of its Creator and I, the observer, have seen it to be so – a fine piece of clock-work where each Particle is Regular and Necessary for its neighbour. Papin, who thinks always in mechanics, might enjoy de Mornay,[3] a Huguenot of the most enlightn'd Notions, who wrote of 'the huge engine of the Skyes' As for Papin's thesis, I am much inclin'd to add to my list of many years past. There may have been 24 in all, but no matter, Papin's design tells me I should add another.

The building of a Self-acting Continuous Engine powered by Steam.

*

8

'today is a Red-Letter Day'

Thomas Newcomen is the son of Elias Newcomen. His mother died when he was an infant and he has been raised by his father's second wife, Alice Trenhale. He has followed his father into business and like his father he is a determined Baptist. He is 32 years old and as yet unmarried.

Thomas Newcomen
Dartmouth, 1695

I use a quill only when business asks but today I call a Red Letter Day [1] well worth the telling for I am benumb'd! My father is a man whose face, as my sisters say, wouldst be the better for a smile. Elias Newcomen goes about the world as if he was a Russian bear let loose from a visiting Trader down at Quayside dock, growling at this and that lyke Methuselah feeling his old bones. Even though I am a grown man whose right arm can smite iron whilst I sing to the Lord, my father can make me tremble, but today he hath softened. He hath confound'd us all. He hath stood unabashed in his Righteousness which seemed a comfort to him and an Astonishment to us.

He hath given his approval for my association with the plumber John Calley and declar'd my business to be the best ironmonger in Dartmouth, not that I hadst much competition he barked, as if I was responsible for the failings of my fellow Guildsmen. He announc'd to all who wouldst listen that John Calley, being Dartmouth born and well respected in the Guilds, wouldst be a good partner and together we should make something of ourselves if we put our minds to it. I was mighty glad Calley was not in sight for, being taken with Surprise, I did not know where to turn as unmanly tears stung my eyes.

My father is such a hard man; I did not know he hadst praise in him. He hath beaten me into the shape he desires since I was a boy so that I might be of service to my Parish. Now he burdens me with Treasure and a Purpose known only in Dreams. This purpose will not make me content with a regular life for I hath in mind Fancies greater than my poor language can express and this day my father hath spoken such if they were his own. You should go about exploring the stories of learn'd men who once favoured steam for its power, he said. It is a blacksmith who will know the speed and distance it can travel and its fierce burning of a man's skin. He looked at me steady and uncloud'd in his features and spoke softly as if I were a child again and, more, his favourite, and said that something must be done to render Mercy to the lives of the miners hereabouts.

*

I cannot sleep for fear that time hath shrunk and I with it. I hath not for many years thought on it or cared to think on it not being worth a memory - but here it is; I may pull at the covers but I am back in my slat bed, a boy rising in the dark to collect coal, my hands black and my eyes running from dust long before my master called. I am to this day wary of the cold for I shiver'd for many months at the edge of the Forge begging for a command that wouldst call me to the warmth. I did not know the warmth of the forge wouldst be my new mother and father and I wouldst long for its protection against cold shivers of air.

Except for the Sabbath my days were days silent of name and it was my dear sisters who travell'd those rough miles to Exeter on my birthdays so I might not suffer entirely the Separation that chafed my heart. I never thought I wouldst be glad of the writing down of thoughts long forgotten but this night they flow thick and easy and form unbidden. It was my great consolation that I could comfort my sisters for I shew'd promise in learning my trade. I know now that since my Master was one of our Baptist Brethren my father hadst no concern to my Mistreatment but he wouldst hath settl'd no such Considerations in my contract. Many apprentices lived in fear of a beating and I was one. When I became a Striker, I rejoiced to swing a heavy Hammer under my master's instructions whilst he held the hot iron steady at the Anvil and tapped with a small Hammer where to strike the iron, declaring me a man never to miss. He taught me to spy the colour of forging heat with its bright yellow-orange when metal becomes workable and told me to trust myne own judgment for I wouldst soon learn by its consistency and shortness under my hammer if my timing was right. Here was the blessing of my servitude; I found in the warmth of my labour a certain Peace.

I hath never spoken of this but it made me love God more. As my hands grew stronger I saw the Purpose in twisting the soft iron for a hundred different uses to make less the Burden of many of my fellow man, however humble the handle, knob or nail. For I am no scholar and my interest lies in the making of things. About this my father's judgment hath prov'd true.

I am long made a tolerable Ironmonger and from this day I am a man of Business with Treasure behind me allowing a new Purpose. My father knows the value of keeping well astern of politics but he is also a man of God, a Puritan under his periwig. The Late John Flavel,[2] an Oxford man who was much put upon for his preaching of our Faith, wouldst not hath resided in Dartmouth and been our tutor if it were not under the Shelter of my father. I can say the Newcomens well know the Balance between losing the love of a King and losing the love of God and I pledge this day to keep that Balance and grow the family's fortunes in good service to our Baptist Brethren.

I will show Calley my drawings and learn what he reckons of my Fantasies and in the morning sing with my sisters, for they hath been my Protectors when I hath been a Lamb far from the Sheepfold.

*

9

'my Paddleboat free of Tide and Wind'

Denis Papin has a degree in medicine from the University of Angers but has devoted his studies to steam power. He is a Fellow of the Royal Society. He is 51 years old and writes from Kessel, Germany and has married a widow with a daughter.

Denis Papin
Kessel
Germany, 1698

    I am hidden away in my foul closet called a study-box but more resembling a monk's sad cell, but still to my liking as no students will knock and seek to burden me with their particular *tragique*(1) which robs them of thinking power. Leibniz's letters lie unanswered and press upon my particular *tragique* which envelopes my waking hours. He tells me I chase the principles of motive power like a Hound on a warm scent.
    And he is right - as always. I have run this Fox into its lair. I have it before me in neat lines on the most substantial paper I could find. It is no boast now I am certain. I have both the Principles and the Design to make an Engine of great power! It irks to bide my time with my head on my paws like any diligent Hound and it may be a long wait. I bemoan the Impasse upon me where my Inventions are far in advance of the means to enact them.
    I will raise myself from my Torpor and write in answer to his correspondence for his friendship is a great Boon to me. There are precious few men with greater minds than Leibniz. He never ceases to praise my Motivation which he calls it my Art of Striving, but has, at one and the same time, described me as a steadfast blood Hound then

a Chicken picking at his brain. Since he is an Eagle flying high as a most distinguished Polymath, I do not see he should be so absorbed by earth-bound animals such as Hounds and Chickens except he is in need of company!

My dear patron, Monsgr. le Landgrave, sought to discover the source of Salt on his Land from his salty springs and I persuaded him that my Engine could move a great Quantity of spring water. I had some success but when the joints and valves of the machine leaked, the Monsignor lost all interest in my work and who could blame him? He told me he could not abide to see his good Thalers splutter and dribble water over his land and turned his back. I swear if no blacksmith can be found to seal such engines our theories will become the sad Whimsys of Madmen.

To that cause, I am working to improve furnace heating with a new bellows design. If we can refine our ores to greater Strength, we will be more in the making of larger cylinders. If it is true that the power of an Atmospheric Engine relies, not on the steam itself, but on the atmospheric pressure inside the cylinder it follows that the larger the cylinder, the stronger the pressure. I am presently observing the degree of heat needed to have an effect with a given quantity of water and the size of cylinder required.

Any man who contemplates the power of steam as I have done will agree that using steam as a pump may be a tawdry Conquest, for Steam power has greater potential than the mere lifting of water! I have shown that steam rises to a higher temperature when under pressure and it serves to reason that the higher pressure, the greater the power. Leibniz agrees and uses the principles of Rarefaction. If I understand him, this means that the expansion of steam leads to a reduction in density as it moves which would be the opposite of compression. Therefore, he believes that the idea to use steam to raise a column of air, *id est*,[2] push a piston up only to leave atmospheric pressure to push it its down again, is limiting the power of steam. He writes, *quod potentia infinita vapori* - the power of Steam is unbounded - but I am left to peck and scrape the funds to shew this to be so - if men would only follow the truth before their eyes. I have demonstrated its power via a working Piston, clear for all to see. What was taken as a Toy by some and a Curiosity by others was a Herald of great possibilities but measured by lack of interest, a Herald calling to deaf men.

My device was a simple process. A 3 Inch diameter tube was filled with steam, as any Cook could do. The steam raised a Piston to its highest point where I hooked it by a fastener. When the fire was removed, the cylinder cooled and the steam condensed. On releasing the fastener, the Piston was drawn quickly down via the Vacuum thus created and this force raised a 60 lb. weight! This may be so, Leibniz wrote at the time, but the force of the engine which is the power stroke, is on the condensation phase rather than the steam phase, and this is its limitation.

Not deterred, I have made a little model of a carriage which is propelled by this force, but I fear that such carriages will not easily run on the pitted and sharp bends of our roads and consider that the smooth surface of canals are more suited to the idea of such an engine. I see no reason why this little engine cannot drive a boat if I can transmit the motion of a Row of Pistons through attachments such as Racks and Pinions to paddle wheels.

It would be a great delight if I could construct a small Paddleboat driven by steam! To guide a boat where the movement was not from man or animal or winde or tide but by another power entirely would be Progress beyond anything hitherto imagined.

In a favourite Reverie, I arrive down the River Leine, passing by its famed verdant banks where I am told splendid Alder trees spread golden shadows and water their toes. I would glide into the heart of Hanover.

I would call, Leibniz, Leibniz, where art thou?

It would cause great Astonishment to my dear friend and his fawning aristocrats to disturb their *pique-nique* (3) under those Alders and see me arrive on my Paddleboat free of Tide and Winde and propelled by the Mystery of Steam.

*

2 July 1698
Letters Patent No 356

Thomas Savery
Raising Water by Means of Fire.

A grant to Thomas Savery of the sole exercise of a new invention for raising of water and occasioning motion to all sorts of mill work by the impellent force of fire, which will be of great use and advantage for drayning mines, serveing townes with water, and for the working of all sorts of mills where they have not the benefit of water nor constant windes.

Original Document: *2 July 1698 Letters Patent No 356 Thomas Savery*

10

'For when I am weak, then I am strong'

Thomas Newcomen
Dartmouth, 1699

I am no stranger to the weariness of a day's labour but it is an entirely different thing to ponder the making of steam power. To this end we hath burnt good candles until dawn John Calley and I, burnt and bruis'd ourselves on bending iron to our will and slept like dead men till noon! It burns us more than we learn by inches instead of feet and at the mercy of our mistakes! There are many who say such fantastical notions are not the business of blacksmiths and plumbers, and more, that our work is merely hot air from our backsides since Captain Savery stole a march on us.

It was bitter news when Calley arrived from a late Exeter coach very flustered and with a face like a gib cat. Do you know of a certain Captain Savery supposedly of these parts, he asked as if I was the cause of his displeasure.

I hath heard my father talk of the Saverys at Modbury, I replied.

That must be the one, Calley almost shouted. He hath chalked us well and truly.

How so, I asked and the answer was not to my lyking.

It was hard to hear that Savery hadst produced a repeating pump by the Expansion and Contraction of steam and harder still to learn he hadst demonstrat'd its power to the Society. Moreover, he hath made great Claims for his Invention and sets its use for mines, townes and mills which is very much our desire. He hath named his engine as *The Miner's Friend* that is, *An Engine to Raise Water by Fire*.[1]

I did not see Calley for days and thought him lost to me but he arriv'd at our workshop most flustered and to my surprise handed me his Bible which he hadst covered in his bag and asked me if I was a man of God. I replied that I believed I was.

Then show me some way forward for I cannot find any, he replied.

I took his Bible and turned to 2 Corinthians 12.9-10 and read *"I will boast all the more gladly of my weaknesses, so that the power of Christ may rest upon me. For the sake of Christ, then, I am content with weaknesses, insults, hardships, persecutions, and calamities. For when I am weak, then I am strong."*

Calley looked at me as many men hath done when I read the word of God. My faith is in my voice and even I can hear it.

When I am weak, then I am strong, he repeated. Yes, I said.

We stood there together Calley and I for some time before he asked what I hadst been about and I told him I hadst been stoking coal for our small boiler.

Is it still your aim is to make an engine which pumps water faster than any known method, be it winde, water or animal, he asked.

Yes, it is, I replied.

Then we inquire to mine Proprietors as to their lyking of Savery's *Miner's Friend*. He may claim all he lykes but it will be they who cast a telling judgment. And so saying Calley put on his work apron and gave me a smile lyke a fox I once saw in the moonlight and I praised the Lord he hadst return'd.

*

11

'to have cast a steam cylinder'

Denis Papin
Kessel, 1704

To be a maker of things never before made and never before realised in the mind of men requires the patience of Job who was, if truth be told, a lucky man not to be an Inventor. But who would preach that the belly of a fish with all its horrors, would be a more restful torment than the fitful trials of iron casting?

    I have travelled north to Veckerhagen so many times in all weathers my horse and I could travel in our sleep and arrive without alarm. It would not surprise the foundry men if I fell of my horse and stood back up, still snoring, my mind in the comfort of my wife's bed in Kessel.

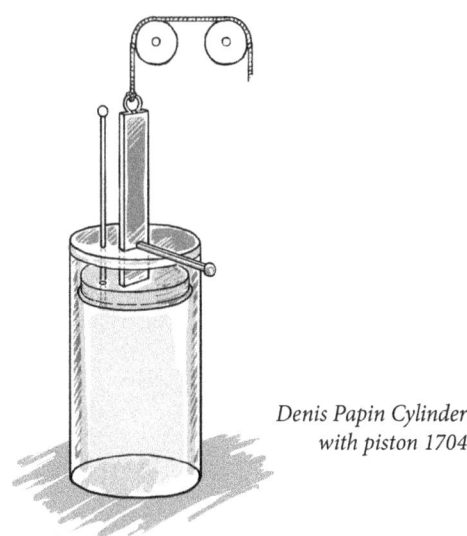

Denis Papin Cylinder with piston 1704

My little boat is now constructed, with the power of steam linked to its paddles and now I need a steam cylinder cast to my dimensions. It will be the first of its kind and so confounds the master smith at Veckerhage.

That I have such an order before the ironworks makes me as popular as a multitude of locusts nesting on the Weser. To join metal, or weld true to the round is beyond most smiths who are used to working on flat planes joined nicely by heating and hammering. They listen respectfully their eyes narrow and appearing to have a suitable gravitas until they laugh and offer me any number of wrought fountain railings, well heads, door knockers and if I was so inclined, a most favourable Grave Cross if I wanted their best work. I laugh with them for they are good fellows whose arms must ache with their endeavours.

Leibnitz sends word for me to persist but he has no need to do so – upon the casting of such a cylinder as my dimensions insist, rests our whole enterprise and all that may come of it, which, when I am set by a comfortable fire with a good draught, seems to be a boundless sea to lands of many riches.

*

12

'Fool's Business'

Thomas Newcomen
Dartmouth, 1705

According to my wife I am about Fool's Business! We hath great bills which earn the wrath of my Bride who is amaz'd at our Mechanics. She stands with her hands on her hips, a sight which sets my heart a thumping, and tells me she married an honest ironmonger and not a Philosopher who wouldst bring Monsters into the world on the back of good money.

She is young and hath more energy than any man could match for she makes of the day a thousand adventures all washed, cleaned, cut, dried, beaten, creamed and pressed and joins me to pray and sing fresh as dairy milk in the morning. She declares she will forgive me for my singed hair but not for frightening her chickens with the Fool's business in my workshop.

If my wife were not such a joyful spirit who makes me laugh I wouldst chastise her for it is only proper that a wife be at the service of her husband in all things. Yet Calley tells me Mistress Newcomen is most popular in her support for his efforts to raise funds. Calley sings the praises of our Engine to Investors and hath persuad'd his neighbours, the well-connected Dottin family, to invest. My nephew Elias hath also persuad'd Cousin Trengove, to become an Investor. Elias hath declared Calley and I to hath the most desirous adventure in Dartmouth and he cannot keep away. He has gladdened our hearts and added to our zest. I cannot bring myself to ask for Treasure but I will tell anyone who will listen that a pump driven by steam will well ease the flooding of mines which is a Song of Woe through the hills of Cornwall.

I do not know (but I pray) if I will be granted Everlasting Happiness but I know I do Christ's work this day. Who of us hath not known men who hath drown'd in the dark and their wives and children left to bend their backs in sorting ore at the mercy of all weathers which hath them so crippled they beg on the commons?

My father, God rest his Soul, spoke of miners as men of stoic Virtue who work in heat beyond Apprehension. He told me once he knew of men so starved of breath they saved their candles to give them air!

Who hath not seen Proprietors deathly pale as they talk of yet another deluge as the earth spews its filth towards them? Who doth not know the heavy cost to lift water either by means of waterwheels and horses, only to watch them thrash about or be chilled to sickness in winter? I will tell all who listen that we cannot hath men robbed of Profits earned by honest effort or our villages starving and our hills crowded with filthy pit lakes.

As for Captain Savery who hath taken the Royal Society by storm with his *Fire Engine*, and choked us all with his Patent, we fear him not! Savery's engine is more used to raise water in the Fountains of Grand Estates by men who can afford its services than bring relief to the flooding of mines. We hath further heard many collierys say it is not worth the trouble as it is slow and expensive and worse, it hath gathered a reputation for explosions. I hath been told horses are more reliable and are generally favoured over Savery. Such reports give us much comfort, no matter that Savery ignores his detractors.

The good Captain hath boldly advertised that his engines were ready for use and might be seen on Wednesday and Saturday afternoons at his workhouse, no less, in Salisbury Court, London, which sees itself as the centre of all things. He hath also given the world a description of his Engine in his book published for this purpose. We found he hath boldly answered those who call it *'a useless project'*. He writes it is such a robust simple engine that he hath boys of fourteen attending and working it, *'to perfection'* He boasts that his engine may raise as much water as two horses working together and since ten or twelve horses must be kept as pairs to relieve those horses, such an engine as his could save the costs of keeping those horses. I hadst never thought of that Logic but by his reasoning it wouldst add a benefit to persuade any mine owners wary of investing in steam.

*

It is well before the waking hour and our rooster cock hath not shaken his comb but I am quill in hand to keep notes of all that hath befallen us, or astounded us so our lives grow beyond all reckoning. My wife still insists we are building a Folly, but she listens well and understands our secrets are to save us from Gossip which thrives on ungracious Ignorance. It is not so much as we are engine-makers which astounds me but that we hath become readers of philosophy rather than our Bibles.

Calley says the purchase of Royal Society papers is worth more than the secrets of Alchemy for turning our labours to gold. He speaks of the Huguenot, Dr. Papin, whose proposal to move a piston within a cylinder seems a snug fit to our thinking. We discuss such papers at length over every meal and ale we share until we laugh and call each other Professor. Calley was bold enough to correspond with Dr. Hooke, the Curator of Experiments, [1] for the Society. We deemed it an honour to receive his reply, albeit brief, given we were Smiths striving above our station. He sought to put us on the path to success as he knew how costly such Experiments were on the Scale we were planning. However, Hooke did not favour Papin's air pump nor did he believe it possible to create an airtight pipe, but he admitted that a speedy vacuum under a Piston could mean our work was done.

I think he means our thinking work is done, I said to Calley.

It is a pity then, jested Calley, that we are Makers not Thinkers for our work wouldst be done and we could go home rich men and take our ease.

But read on, he added, read on.

It is my fervent wish, wrote Hooke, that you should not be led down rabbit burrows.

I recall the silence as we considered the honourable Philosopher's advice. I may be a man well covered in years but I do not think I am of the age of Senility not to know a rabbit burrow when I see one, Calley declared. By the shadow across Calley's face I was thankful Hooke hadst not taken it upon himself to visit for his Consultation.

We might not be University men, I told Calley, but every consideration hath led us to one conclusion, power will be gained by using a steam Cylinder containing a Piston.

That is the way, replied Calley, steam will force up the Piston and condensing steam will give us a vacuum and gravity will draw down the Piston. He leant forward and stroked the side of our small clay cylinder lyke he wouldst a prize Heifer and I thought on the instant he might lead me on a jig round our workshop!

*

Mistress Newcomen hath chided us for making her ill with our endless tilly-tally over Captain Savery which she swore made her grind her teeth so she might one day spit them into our soup. She begged us observe what device Savery hadst made to overtake us and so we did, for, in truth, we had not thought to do so.

Much to my discomfort my dear wife stood by, her arms folded akimbo lyke some carved creature braving the winds for sailors. I could not chastise her for she hadst a light in her eyes I hadst not seen before. Calley and I bent our heads lyke new apprentices and shewed how we could fill our vessels with steam. We used cold water from a small cistern to cause condensation and so make a vacuum. Water was forced up our ascending pipe to fill the vacuum but it was necessary to repeat the action. Savery hadst placed a Valve to stop the water from descending (which in every county of the Kingdom it is wont to do) whilst more steam built up force in the boiler. On being let into the vessels again, the steam pressure caused the water to raise a second Valve and ascend to a height where it could be released into a drainage pit. This increase of steam pressure, we noted, was the probable cause of explosions.

Mistress Newcomen stood for a moment and looked at Calley and I.

It is too slow and hath too many actions, she said, and went back to her housekeeping. I did not know Calley could laugh but he did and slapped his thigh with it and I laughed with him. We hath been so told, said Calley and we laughed again lyke boys free to run.

Moreover, we hath learnt that Savery's system needs to be placed no more than 30 feet above the water level which limits its operation. Our judgment so falls against Savery's engine, not only because of the difficulty of making the boiler joints tight, but also that a great part of the power of the steam is lost. Nevertheless, to the greater good, Savery hath shewn that a furnace, a boiler, pipes, cocks, valves and a cistern are all that is needed to harness the power of a vacuum using steam. None of these devices are strangers to us.

Happily, Savery's engine will spur us forward for we know how to better it. We will not use a vacuum to pull the water up, our vacuum will have atmospheric pressure to pull a piston down. We will transfer this movement to a pump rod. As I know my father wouldst suggest, we should make enquiries among our Brethren in the North, for the height of the iron cylinder we will need (taller than any man we know) are beyond our capabilities.

We must endure a lesson in Patience that we are loath to suffer for Savery hath cornered the field by his Patent, which is grandly titled *The Fire Engine Act* and is known from one end of the Kingdom to the other. It is a wonder that he hath not sued his Rivals for our mere Thoughts, for if they could be read by any man, they wouldst be seen to be well advanced from Savery's invention. But a man who hath no Patent is a man who hath no Business and those of us who think on the possibilities of steam wait lyke so many sheep at the will of Savery's Patent dog.

*

13

'this cursed coach over these cursed roads'

Denis Papin
The Dover Road, 1707

If I was not grateful to be carrying my dreams in my Luggage and my body in one piece, I would curse this long day crossing la Manche [1] where I have been held bobbing on unfriendly seas on the whim of a Captain not knowing his tides, and now my Arse is to be bruised like a whipped schoolboy in this cursed coach over these cursed roads.

    I am lucky that the mob of Munden [2] Ferrymen who seized and destroyed my poor paddlewheel boat some days past, did not Seize and Destroy me, for they did a passing good rehearsal of desire to Murder. I had made quick passage to the Weser River but these fine fellows were waiting on the authority of their Guild to stop my experiment and took delight in my quavering retreat, for I was but one and they were many. They had decided that no such paddleboat and no such Ferryman as I presented would trespass upon their river and stood with cudgels to make it so. Had they known of the success of my fire engine to drive my paddles against the current I may well have more bruises to complain their treatment as I jolt about in this shabby coach. The coach master assures us with such clement weather we will make London by the late morrow where a fine meal awaits us at The Spread Eagle. He has said this so often I am sure that on arrival we will meet whichever member of his family is the Inn Keeper. I am back in the land of meat and potatoes, and a home-grown Jenever so gut-burning a brutish swill I could use to fuel my engine if I could find a way!

    I write this at my lap desk that has impressed my travelling companions into a wary silence. Perhaps they think me mad but it is better occupation than the morbid shanty they sang on the crossing. Besides, by the time we see Canterbury's Spire I predict my present humour will have melted and I will feel again a fondness for these green fields and black-faced sheep as if it were my native land, for in truth England has offered me much comfort since my King has seen fit to declare good Frenchmen who Worship Christ enemies of the Catholic State. His prerogative and power have been exceedingly fortified of late and we Huguenots have been the means to his greater licence.

If my mood is still bleak I can procure myself some Jenever tonight and drink myself into happy Oblivion in some Canterbury Inn. At least to release myself from the opprobrium of my companions who suffer my smell and fashion as if I am a Savage from beyond far seas.

As I am a returning Emigré more hopeful than certain of my standing among the Luminaries of London, I will not indicate to them that their smell offends me also, in equal proportion to a loaded Privy. It would be of no interest to my companions but I do not arrive empty-handed to impose my keep upon myne Hosts like an unwanted relative. I bring to London the drawings of two workable steam engines, one from my work in Marburg University and one from Kassel on the banks of the Fulda where my paddlewheel boat began its fated journey.

In my luggage, which presently thumps its weight happily upon the roof above us, are the books and drawings worked on by myself and my colleague Leibniz whom I have just left these days past in his workshop in Kassel. With Leibniz's connections, I induced the foundry at Veckerhagen to cast me a cylinder by which, I believe, I have had more success than any man to present times. I do not have the funds to transport such a cylinder and have left it under his care.

My designs are an advance of any notions from Captain Savery of Devon, as Leibniz has confirmed. I plan to demonstrate this to the Court and to the Royal Society so that immediate steps can be made to build such machines for the benefit of shipping which I hope to demonstrate on the Thames seaport. The effect of steam power on ships will hasten a New Age beyond any hitherto seen! A Ship to saile with all Windes, I shiver to think on it!

Leibniz is a genius to rival any Englishman and together we should enchant any Society Fellow who applies an impartial mind to our Designs. It maybe years since I first wrote of the use of steam to drive a piston, but it matters not. Steam may be a new land, nay, an entire continent, but I have mapped it and spied a landfall. It is a landfall with the same Principles as when I demonstrated to the Society how it could cook otherwise improbable foods and bones etcetera in my *Digester*. I was then a hopeful man in my prime and hope still abounds in my bones, even now I am, according to my wife, become a crusty curmudgeon as time weathers me.

Perhaps she could ask the learned doctors at Edinburgh to draw a table where the amount of pain received on the nether regions on a coach seat can determine years left on this Earth for if that were the case I should expect an early death.

*

14

'We hath the means via licence'

Thomas Newcomen
Dartmouth, 1707

Calley calls it Providence and I call it God's hand, but we happily agree on our great fortune. We hath a working model and now we hath the means via Licence to move forward! Some time ago Captain Savery visited Dartmouth on his duties to inspect our hostel for Sick and Hurt Seamen. It was an Opportunity that even Mistress Newcomen wouldst scold me hadst I missed. I found him at the end of the day tired and worn by his task and so happily diverted to a meal and a discourse on a subject dear to his heart. He was no stranger to the name Newcomen and hadst heard of our endeavours. I put it to him that steam was our Mutual Friend. We are both men with calloused hands and his eyes shone when I spoke of my Brethren Iron-makers in the North and my knowledge as a smith.

We were of one mind in the importance of our project to Industry and Prosperity, and I persuad'd him of the mutual benefit of a Patent agreement between us. He was especially of a mind to join us when he saw the progress we hadst made. He professed that our designs were similar to his own and could hath been made by the same man.

Business hath made me a Diplomat and I nodded in agreement and did not dwell on unnecessary Differences when Similarities would win the day.

He glowed with pleasure and declared that his Principles of Motion hadst reached their limit and new minds were needed to carry forth the quest for such a power as steam offered us. As I write, it is a great satisfaction to know we are united under his Patent which, due to its extension, lasts well until 1733, time enough to hath a hundred working engines if Fortune and God's will favour us.

As our engine is a Giant of wood and iron, no man can steal our design and hope the world will not notice. Savery is full of assurances he will set his lawyers as the Hunting Hounds they are upon any such miscreant who is in breach of the Patent. We are in agreement that we should be so protected for we hath made advances due to our own Diligence and Expense.

I am still Amaz'd that I can write of such things as if they are a mere Inventory of Tools to be listed for Business when I should be on my knees in Gratitude! We hath a way to make the action of the Piston continuous, and this action will be at a regular rhythm for we hath put the Principles of Balance, Gravity and Weight to work in Harmony for the good of the Whole -a rocking Beam with a Piston on chains at one end and a long Pump Rod on the other. If we are correct, we hath overcome Savery's limitations, for the suction of our pump is on the upward stroke and will force water from a greater depth. That our steam must condense without delay now plagues our minds. We plan to place the boiler under our cylinder so as to directly supply steam and this will help its timely condensation.

In my daily prayers, I ask my Saviour for His Blessing on our Endeavours. Savery advised me to do so. He is complimentary of our Advances but hath an Inclination to keep his distance. He described to me the years when Experiments discouraged him so that he wished his Torment over for he feared for his life. He confessed that many of his engines could not be brought to answer and that several exploded and rent the Whole to pieces. In particular he told of his first engine, assembled for some reason he could not recall, in a Potter's house in Lambeth south of the Thames, where upon heating, the engine forced its way through the roof and lodged in the Tiles where it smouldered and threatened fire to ruin all.

I was set to laugh at his description until I saw his lasting Mortification at the horror this caused in the Spectators present so many years ago. Further, he lamented that sometimes he fancied himself Ulysses tied to the mast while only he heard the call of the Steam Sirens and his workmen did not, for only he understood what they were about!

I am a now made Pastor to my community, a singular honour to my faith and family. I hold no fears that such torment will attach itself to me, for I hath about me my Baptist Brethren, all God's good men.

*

15

'a Force well beyond the imaginings of my critics'

Denis Papin
London, 1708

London is much changed from the time as a young man I walked its alleys and by-ways watching the restoration works after the Great Fire,(1) when preachers told the City that it had been Chastised by God and men were eternally grateful for their Deliverance. I was barely able to speak a word of their blunt tongue, my only friend a Letter of Introduction in the hand of Christiaan Huygens, my dear professor and mentor, long since departed. It was entirely upon his Recommendation that Secretary Oldenburg met with me and befriended me with lodgings and employment with the Society.

Today I walked those same cobbles among the Bustle and Noise as a happy ghost, returned on my own recommendation with Designs in my own Hand, able to converse in English with my back still straight and hope in my heart. I smile to be an Outsider still, with shoe buckles too large and a periwig too bold but I must confess not one man gave me a scolding glance, so much has the city changed. It is a motley collection of houses thrown together to give every man and his dog a home but it has theatres, clubs and coffee houses. It may boast of its Architecture and Music and the great advancements of its learning but it is still bewitched by Trade and Business and the getting of money every minute of the day. Not a man will admit he sits among the usual *compost* of animal dung and rotting meat and bears his own stink until the rain washes it into some other fool's cesspit. If I had a coin spare I would employ a street urchin to sweep a crossing for me and make of London a passable city. But if I drag my eyes from the muck of their streets I see their builders mark the Glory of God. The Cathedral of St Pauls on Ludgate Hill at last shows its dome and spreads its majesty over all. And to my relief it is now possible to have bread made in the bakery of a fellow Huguenot and taste a pastry that will, of its own sweet being, bring a tear of nostalgia to an exiled soul.

It is a sad Truth that we appear to Londoners as Omnipresent throughout their city, claiming with our cleverness much of a dozen trade esp. of lace-makers, and silk-weavers to the great Resentment of many. They may buy our lace but they see us as Interlopers not as men of Misfortune, huddled together in Spitalfields, keeping to ourselves with the comfort of the familiar and licking the scars of Tyranny.

I have found lodgings in Westminster which is an hour's good walk from my countrymen if I am in want of society, but it was all I could obtain. That my friend and mentor, the great chymist Robert Boyle, has long since departed this earth has left me more bereft of Company than I imagined and the City presents a blank face without his patronage. I left a copy of my thesis for presentation to The Society which I titled *Proposition by Dr. Papin, concerning a new invented boat to be rowed by oars, mover with heat.* and asked for a sum of £15 to undertake the building of such a boat. The Council referred my design to Captain Savery, which was as good as pissing on my shoes and blaming a street cur. The proclamation of the *Fire Engine Act* of '99 had made Savery Master of Steam in the eyes of President Newton and Secretary Sloane who command all other eyes in the Society.

*

There was no surprise in Savery's reply, which was today delivered, but it pained me nonetheless for it was tailored swift to undo me. It turned to Ash many years of work and spurned the great mind of Leibniz, who deserves more Hospitality towards his Intelligence. Savery damned me outright. He would not support my cylinder and piston design and declared it too prone to Friction. It mattered not to him or the Society that I had led the way on the power of steam with my *Digester*, nor that I had worked diligently with Robert Boyle and presented many papers of Merit, nor that I had been declared a Fellow, nor given my best years to the advancement of Knowledge in the hope of easing the Labour of man and advancing general Prosperity.

It occurs to me that I have been a Convenient Émigré and done these Englishmen the service of easing their consciences for the Protestant cause. It is now my foolishness entirely that I thought all Fellows were as honourable as Boyle. Of course, no man will see fit to take up my Application without fear or favour, for they sit in Sir Newton's Shade as if he was the Sun King himself!

If I had not felt the ease of the Paddlewheel boat against the current right through my legs and into my belly, and thrilled at its promise as it pulled through the river's flow, I would have surrendered all and called Obscurity to take my name and blow me away on the next Winde. But I did stand on my little boat and feel the future promise of Steam greater than a hundred oarsmen. Had Fate favoured me, I would have parted the waters of the Weser and made my way into the salty waves of the North Sea. It is so clear in my mind I cannot believe others cannot be brought to see it.

If he were here, dear Boyle would, I am certain, seek to cheer me on. There was no better man in all weathers. I do miss his storys, for he was given to Fancy as bright as his Sunday smile. If it were not the storie of the Barbarous Fens of Ireland, it would be the Sublime Beauty of the Italian landscape. I must have heard the storie of him as barely out of childhood meeting the great Stare-gazer Galileo while wandering the hills of Florence a hundred times without tiring of it. That Robert would embellish his tale with various flavours and asides, differing with every telling, helped convince his listener of the freshness of his memories. It did not Discolour his telling that the death of the Great Astronomer fell on the very day he was supposedly in great Discourse with him, for very few know of this unhappy coincidence!

It gives me comfort to recall such days in his household when the world was not such a serious place and men more brothers with open hearts. These moderne Englishmen are pale suckling cousins to Boyle but it is to their credit they have honoured his endowment for public lectures. This week the Theologian John Turner will defend Christianity and the Redemption it offers in his Boyle lecture but I may not have it in my heart to go. (2)

This night I will walk to Spitalfields and indulge myself in a dinner not baked to a crisp and some wine and cheese which speak of home. But before I turn tail and let a rich wine dull my senses, I will write to the President of this Royal Society so proudly in affirmation of Reason & Experiment, and beg a private audience to plead my case. I do not think he will want it known that he refused such a request from a Fellow of many years standing.

*

16

'the President sat, a mighty Earth round which
the Sun and all the Planets must run'

Denis Papin
London, 1708

I have this day presented my case to Sir Isaac Newton, our Honourable President, a gentleman of Science most gracious with his time but little else!

When I entered the President's hall he waved a greeting and bade me sit to his right at a long table where, it seemed, the Council held their deliberations from the row of pipe stains. I was grateful that there were refreshments by way of a Jerez Dulce sherry to ease my dryness, but the English habit of drinking such a beverage in the afternoon is beyond understanding. Newton greeted me kindly enough and remarked that it was good to see me returned to London, for it was growing to be the centre of all significant Discoveries and was already the envy of the world of Natural Philosophy.

I did not take up this assertion as I knew any Defence of the Academies of Paris and Berlin would stir his discontent towards his enemies on the Continent. He seemed pleased with my silence, and sat tapping his fingers on the sides of a mahogany chair carved very much in *le style francais*. He was indeed a Longshanks as the English say, stretching his legs before him and musing momentarily before he began. He made reference to the time I had demonstrated to the Society.

It was over twenty years ago, and we were both much younger and more vigorous men, were we not, he said and smiled as a man wistful for past glories. I nodded because it was true that we were now men who had seen our three score years but I refrained from entering a Defence of my Vigour, currently pulsing hot in my veins.

Our dear Robert Boyle, God Rest his Soul, was your mentor, was he not?

I said he was.

And it was your *Digester*, was it not?

It was Sir Isaac, and as I recall it created a sensation, I replied.

Indeed, it did Monsieur Papin. It demonstrated sound principles but where has it gone since, I wonder?

There was no denying very few people had been willing to make use of my cooking pot but I kept my voice as occasional as Newton's.

In the case of my *Digester* London craftsmen were not capable of realising its potential, I answered.

There was another silence as Newton was absorbing my rude diplomacy regarding English craftsmen and I was reminded that an Englishman's sensitivities well outweigh a bold truth.

But if I may account for my poor Invention, I said, and not to be bound by his silence, I continued.

Since we work by process of demonstration, we must always allow for surprising fruits of our labour. The pressure relief valve I designed for the safety of the whole contained a Piston which I held in place by a Weight that would move up and release steam when the pressure grew too high. This release would ensure the Weight fell and closed the valve.

Yes, I do remember, Newton nodded and smiled. Most ingenious.

Indeed, Sir, it was this process which gave me the idea of using a Piston in a cylinder filled with steam pressure. I have demonstrated that principle in operation. That is, Sir, the wondrous beauty of Experimentation to which I am bound, come what may.

My words raced and threatened to topple over each other for I shook inside as I heard my own pleading. I had reminded him that I held the view of the Heretic against the orthodoxy of the day as Captain Savery did not use a Piston but atmosphere as pressure alone once steam had created the necessary vacuum. Myne did not.

But that is what Captain Savery has criticised of your present proposal, has he not?

Newton leaned forward, seemingly anxious for my reply as if I could alleviate his fears.

Yes, Sir Isaac, Savery has been averse to my design on several grounds which I suggest should be settled by practical application one Engine, being myne against the Other, being Savery's. Do we not work for the better advancement of Knowledge and the Publick Good, I asked, hearing my voice rise as the pounding of my heart added to its force.

Yes, yes, indeed we do. But the Society has before it Savery's engine already made and you say London craftsmen are not up to the task of your Digester let alone a dangerous steam pressure Engine.

Newton leaned back as if he had made a telling point and thought the matter settled.

As my proposal states, I replied, I plan to have the parts for the Engine made in Kassel where I have already completed a successful version of the Engine.

In Kassel in Hesse, Newton asked, leaned further back in his grand chair.

Yes, His Highness Charles, Landgrave of Hesse has paid for my experiments from the same design.

Newton remained silent for a time as if he were absorbing the arguments I had presented. Then he spoke.

In the many years since a King, no less, granted our Charter and we have done marvels in the advancement of knowledge, but such Progress comes at a Cost. He paused and looked at me over the tips of his long fingers pressed together as if at any moment he might pray, and then he continued.

Fellows have fallen Prey to the forces of the Market or spent their Dues on their own Experiments, all to the detriment of the Society, and now we are told we must find new lodgings for our activities as the College rooms are no longer for hire. Our resources amount to a very thin gruel at present, Monsieur Papin and I must ask you to understand that in our present position your proposal is far too costly for us to entertain.

There was a silence which grew while I choked back disbelief that the Great Newton would, in the final act, take Savery's poor Engine over myne. I saw that the President sat in Defiance of his own Calculations; a mighty Earth round which the Sun and all the Planets must run, and that Savery was one of those Planets and I was not.

The size of my chagrin could be measured by the amount of foul Sherry I poured down my throat to stop me flinging curses at Newton, the Society and Englishmen generally. Newton rose and sipped the very last of his Jerez Dulce. I expected him to lick his lips like a hunting Stoat as he finished speaking but he bent his head and wiped his mouth with a kerchief as grey as his face. Monsieur Doctor Papin, that is the end of our particular matter. The Society prays you do not leave London and as a valued Fellow, we look forward to other ideas you may present to us.

We cannot offer you employment at present but please keep in correspondence with Secretary Sloane. You may indeed find work with Captain Savery.

Newton's words echoed in my ears as I made my way down Fetter Lane into the gloom of twilight and its dangers. He is a great genius for his words struck my heart greater than any *electricity* known to man.

My pennies did not amount to a Chair home, even less the Fare to Calais. I could not curse God for I had great need of Him, both on this gloomy night that I might arrive at my lodgings unmolested for the money I did not have, and for the days ahead when I must strive to resurrect my Fortunes in this stinking city where no man calls me Friend.

*

17

'It breathes like a living thing'

Thomas Newcomen
Coneygree Colliery
Dudley Castle, 1712

There was never a sky more blue than this morning at Coneygree. I confess I hadst given no thought as to the Occasion being wholly taken up with our preparations and was surprised by the crowd which gathered apace as if Dudley's men hadst promised them a feast for all. Miners brought their families and gathered round with expectant faces and knowledgeable predictions. The men flaunted long coats and the women wore hats the best Staffordshire could offer. Others came out of Deference to the Baron and even more by the Curiosity in their natures. I was grateful that my brother's son, another Elias in the family, and John Calley were by my side, for I could not hath stood alone on such a day in front of so many. And most strangely, I wish'd my father with me, to bluster about and find some small fault to temper our Souls.

On my signal, the Fireman was quick about his work and with a long moan of sound the piston moved as we hadst devised, lifting a weight of water never before known anywhere in our kingdom or any other. The engine made 12 strokes a minute and each stroke lifted 10 gallons through 51 yards, that is, from a depth of 156 feet! The Managers of the Colliery declared the day a great Success and boasted of Prosperity for all. We were forthwith entertained at Dudley Castle as the guests of Edward, the Lord Dudley, and his mother. His Managers hadst assured Edward, who was all of eight years, that our Engine was a Noble Beast whereupon he congratulated us and asked the most confounding questions.

I hath sat to write to our partner Thomas Savery but cannot for I am utterly confounded. I would hath John Flavel's counsel tonight, as if he were not long dead, and beg his living body fly from its Heavenly Paradise and sit with me in my Agitation. He was a man of great faith and I would hath his hands upon me. It was he who told me the ancients described *Necessity* as the *Mother of Invention*, and set my mind racing with the novelty of the words themselves. Long before I knew what *Invention* might be, I determined I would associate myself some day with such a thing, for *Necessity* itself seemed a great Force, not to be denied. But now I am in Disputation with myself concerning the wisdom of such thoughts. That I should be in such a humour on such a momentous Occasion for our Enterprise causes me to doubt my own Balance of Mind.

It seems a thousand years long past when I stood on raised ground this morning, the better to be heard with Calley and Elias by my side. There was a hush when I began to speak for our engine was a most Uncommon Sight and many in the crowd stood silent not knowing what to make of what was before them. Many hadst never thought it possible men could build such a Contraption or find a use for it.

The engine straddled a dark Pit with its brick chimney at 30 feet and its separate engine house almost as high. At either end of a long Beam, the crowd could see iron rods, one connected to our Cylinder and one connected to the Pump rods. I pointed to the pump end of the beam and said they could all see it was heavier and would account for the Piston rising as the pump falls. I explained that the Beam would move by use of atmospheric pressure to push down the Piston. Since many would not know of atmospheric pressure I said that such pressure was a gift of Nature where Nature herself finds a Vacuum unnatural. I further said that by filling a Cylinder with steam and condensing it rapidly by use of water, we hath created the necessary pressure to compel the Piston's movement.

Thomas Newcomen Steam engine 1712

I did not want any man to say we were not following God's gifts in His Creation and when I stopped my address there was a general mummer of approval although I could not say if any man hadst understood a word for all their nodding. So there, in the morning light with the sun about to make a golden day, the first thud and hiss of movement astonished all who witnessed it into silence! It was Coneygree's manager who applauded for he knew what Treasure it would make him. I thanked God the crowd applauded with him and some cheered at the wonder before them for I do believe they knew not what else to do.

I made to address the crowd further for I hadst enjoyed their listening, when words were taken from me and I hesitate to write what follows, for I wonder if the words, once written, will rise from the page to haunt me for this is what befell me.

When the machine moved of itself and breathed lyke a living thing I stood lyke a man astounded as if I, myself, hadst never before thought its movement possible! Further, I saw it stood Proud and Apart, rumbling as distant drums might, but deeply, seemingly from the bowels of the Earth. Then it sighed its steam sigh, wet and warm and muffled as a savage Beast might purr, and began again, a repeating pattern, steady and true.

    I stood transfixed by that Noise, not startled or repelled by its rude power but drawn in. I step'd forward as if to inspect the movement of the Engine and encountered the smell of heat and oil on iron and brass, so familiar and yet fresh upon my senses. There was a great warmth that reach'd through my bones as if I hadst suffered cold. And, in truth, I stood lyke bothe a man and a child before a new Hearth never before known on Earth and I knew it to be so. I dared not go further but yet longed to place my hand on its mighty Cylinder, for I saw we hadst harnessed a Beast that would work for many years and Lo, I would be its Master and hath such power at my command.

    We hadst worked until we could not stand, learn'd to bear our burns and run for shelter as cylinders tore their seams and steam values failed, but today I stood close without Trepidation, listening to the rumbling motion before me and the regular movement of the Beam's motion to the Piston rod, back and up and back again. I turned away only afeared it may speak, and more, that I may comprehend its words and be bound forever as if I hadst created a living thing to worship against God's Commandment. If not its Master, I might become its Servant and bring upon my people the same terrible rebuke placed on the Israelites for their idols.

    I am hot and full of shivers and can only bend my knees to pray. Rev. Flavel preached that men must work in the service of God. He would remind me of the true purpose of this Engine and the Goodness it will bring. I would hath him cast his loving kindness upon me and place his hands upon my head, the crown of my skull encompassed lyke a midwife's touch and my heart steady as if I was born anew with his Blessing. It is surely true that we are God's Creation and that he hath given us talents which we should employ but today I learned that the making of something and the living with it are two different things.

*

18

'I am in a sad case'

Denis Papin
London, 1712

My return to London has not been as fruitful as my dreams desired, and a man who lives with broken dreams is not a man long for this Earth as wise man once said tho' I cannot recall who. I have been at a low ebb and my health slips away as if it is of a mind to follow my fortunes. There is no man worthy of calling Colleague and no man who has shewn interest in my Proposals. I have watched this Metropolis rise as a Phoenix from stinking ash and stinking marsh and my hopes rested in its new sense of Purpose and Industry but London preserves its Industry for its own. My letters to Secretary Sloane receive desultory replies in a flourishing hand as if his graceful swirls should serve as my comfort. In particular I have drawn his attention to the Society's failure to register under my name a number of my inventions presented to them. I have no funds for the hiring of a Lawyer from the Inns of Court to aid my Complaints, so my persistence is a rubbing sore to irritate my enemies rather than Salve my Reputation.

    I am more often these days at Narrow Street or the Old Swan(1) where I can engage idle Watermen in conversation to pass the time of day. They sun themselves and share their tobacco on stairs above the tide. They are a true Fraternity of men whom, were I younger, I would happily join. They neither mock my accent nor my ideas for they tell me they ferry the Tribes of the world across the river and boast they have become *loqui linguis* (2) as happy parrots to the learn'd men they carry up to Oxford.

    They do not want for understanding of the power of Water and the hardness of their work against a running tide. I talk to them of how water on its downward rush to join Mother Sea has served us well since Christ was a boy by milling our flour and grinding our ore. They work upon a great natural force and are curious to contemplate my theories that Steam will turn out to be its Cousin if contained and controlled. They laugh kindly at the idea of my Paddlewheel boat but promise me that were I to provide them with such a boat, they would ferry me across Free for Eternity.

Despite their companionship, I grow more listless by the day and even as it was salt in my throat, I have written to Sloane to thank him for the paltry £10 the Society has seen fit to grant me in compensation for my efforts. Notwithstanding, I am in a sad case. It is a long while since I left my wife and her daughter in Kassel with hopes of calling them to London once I had the funds to support them, and it is a longer while since I fell on my knees and asked Almighty God for my salvation.

*

19

'I am come from Coneygree for the engine has stopped'

Thomas Newcomen
Birmingham, 1712

I hath learned a great deal about Business from the men of Birmingham. They are most hot for Treasure and sharp about it. It is not to my lyking but seemingly a necessary evil much the same as gutting fish under a hot sun. Our success hath left us with more work than relief from labour for our Steam Pump is in need of many brothers to satisfy the needs of mine owners. I hath found Failure and Success children of the same family for they both require attention usually of an immediate nature, compelled either by Duty or Demand. I hadst been meditating on our future production and the need for the formation a Company to give gravitas to our endeavours when John Calley came knocking at the door of my lodgings.

Tom, he called, his voice all aquiver. I am come from Coneygree for the engine hath stopped and I cannot make head nor tail of it.

To hear panick in Calley was unusual and caused a thrill to quicken my pulse.

Hath it has stopped entirely, I asked.

Entirely, he replied and stood at the door as if I might provide him with the solution poste haste. He hadst a chaise waiting.

I would reckon on it being a problem in the cylinder, I proposed, gathering my boots and coat and calling to Elias to come on the instant.

There are men standing about, Calley said, his face burning with his Mortification. I hath never seen him look defeated, not in our darkest days but now he stood like a boy chastised.

They were cursing us and deciding whether to send for the owners, he reported. Hadst I not been a Christian Pastor I would hath indicated to Calley my opinion of the Coneygree miners but I bit my tongue and prayed for God's Grace in the hours ahead for I knew the losses they would be facing, even by the hour! We were barely three miles from the Dudley Estate when dark clouds became rain, making the way more a muddy rutted lane between hedgerows than a road.

We would be quicker walking Calley offered, his face grim with distaste. I cannot comprehend, he added bitterly, how we Englishmen can make great advances and yet cannot fathom how to make a decent road and are forever bouncing along jumble-gut lane!

As for myself, since panick hadst been known to attract inclement weather, I would rather seek to balance my thoughts than plough the mud around us. I told Calley the world would agree with him but begged him to stay in the chaise. I could not bear the thought of him falling broken and bleeding into some crevice or ditch concealed by the gloom of the rain. As we churned and slid I thought that a sensible man would put his funds into waterways rather than steam engines. They would at least offer a reliable congress for any poor trader who sets himself to travel.

The discomfort of the road was outdone by our arrival at Coneygree where we were greeted with whistles and calls of encouragement bordering on jeers. I could not countenance the thought of retiring and using the observations of models to find the problem.

The Engine stood still before us, waiting lyke a poore pilgrim for a balm to cure all. Between autumn showers Calley and I inspected its cold cylinder and, hadst Calley not been by my side, I know I would hath conversed in great seriousness with the trappings of iron, brass and brick before me. Instead I went to the miners and sought their foreman and was told the ground bailiff, as he was known in these parts, was in Birmingham on business but I should talk to Mr. Turner. He came towards me accompanied by more than a handful of miners, all strong dark fellows with blank faces. I lifted my hat in greeting and asked if they might tell me what they hadst heard before the Engine stopped. They wiped their hands and touched their forelocks. Turner looked on me surprised.

You'll be asking for some help, then, he said and I could not take offence for his face hadst not seen a smile for many years and I did not think he would hath a gibe about him.

I am asking if you can tell us what hast occurred, I replied, my voice as steady as I could make it.

We hath been sat about for the best part of two days, he replied and his men muttered and shuffled about.

I did not shift my gaze from Mr. Turner.

You will hath some considerations then, I said, for you must know the engine well.

At that he nodded and told me they hadst heard a sound lyke a sniffling dog or a sneezing old man and the engine was definitely short of a shilling.

He reckon'd, if I would not mind them sayin,' it was a problem with air. The men about him nodded in agreement.

They were blunt Midlanders but I hadst no grounds for complaint for I reckon'd they were right. It seemed a problem with air, though I hadst no idea what. I thanked them for their observations which I greatly valued and they became almost apologetic and beckoned to Calley and Elias and waved us into a rough shed. Awaiting us was hot sweet tea, strong cheese and some good beef sent down to us complements of young Edward Dudley's men. I was grateful that I would not have to converse with Mr. Turner again for he ordered the men to pack their kits. They left with Hoorahs which we took in good humour.

We unfurled the drawings of our Engine and regarded the familiar connections as sudden Strangers bent on our Annoyance. The fate of our whole enterprise hung now on our Deliberations and, not being Philosophers of Mechanics we could only do as we hadst done so many times, work from observation and questioned what might be the cause.

Where could the air come from? We hadst struggled to make the cylinder as airtight as possible and hadst allowed that the piston be marginally smaller than the cylinder and sealed the gap with a ring of wet leather. If a deep meditation on calculations known to us could hath offered insight we would hath solved our problem that afternoon.

The day hadst long passed as we set out for our beds predicting for ourselves a sleepless night, when a possible solution came to me. I hath often thought that the gentle gait of a tired horse in a chaise can be as relaxing as any Public House brew and while drifting off onto other matters closer to Nature such as the need for a long piss, it came to me that water contains air. Could it be that when the water became steam air flew with it, I said out loud waking my companions who stared lyke startled hares before an unexpected fox.

If this notion be true, I exclaimed, we could well be cursed with too much air.

If your notion be true, said Elias, the extra air could not be condensed by the water spray and will by constant repetition build up. Calley looked most alarmed. Then our engine is more *winde-logged* than an inn-keeper with a fat belly!

We were all in merry spirits by the time we were safely fed and watered and sat by the warmth of a good fire. Calley and Elias were the best of friends when I retired weary to exhaustion from the day's discouragements. I left them muttering to each other about possible solutions in the company of a good port and it crossed my mind it would be a miracle if God led them to a solution in such company.

*

The engine lay idle at Coneygree for many days and I tasted the Mortification that comes with such Arrogance that believes all Knowledge hath been acquired for Perfection to be assured. Elias and Calley declared the solution to be in a type of release valve called a 'Snifting Clack [1] that would allow gasses to escape the cylinder.

Such a valve would open briefly when the steam rushed into the cylinder so the air may escape at the same time. The relief when the new snifting clack answered nicely was greater than the relief of watching the first movement of our beam so many months ago. Some Wag hadst named this valve for its sound, lyke a Man snifting with a Cold, which, in our bleak days at the mine, we were lucky not to catch and be abed with Death's whisper in our ears.

*

'London Gazette' for 11–14 Aug. 1716
Advertisement

'Whereas the invention for raising water by the impellant force of fire, authorised by parliament, is lately brought to the greatest perfection, and all sorts of mines, &c., may be thereby drained, and water raised to any height with more ease and less charge than by the other methods hitherto used, as is sufficiently demonstrated by diverse engines of this invention now at work in the several counties of Stafford, Warwick, Cornwall, and Flint. These are, therefore, to give notice that if any person shall be desirous to treat with the proprietors for such engines, attendance will be given for that purpose every Wednesday at the Sword Blade Coffee House in Birchin Lane, London.'

*

Original Document: Advertisement: *London Gazette'* for 11–14 Aug. 1716 Newcomen and Calley's company, *Proprietors of the Invention for Raising Water by Fire,*

20

'Proprietors of the Invention for Raising Water by Fire'

Thomas Newcomen
London, 1719

I hath spent as many hours making to hob and nob with strangers over a coffee cup these last years as I hath on my knees to the Lord yet He hath seen fit to reward our endeavours. There were those who called us Schemers and turned their backs as we were Inventors of mechanical matters not worthy of consideration. And still, the Lord hath Blessed us. These days I am conversing with Schemers of all colours, who flash jewelled snuff-boxes and lounge in silver-buttoned frock coats leaving us plain Devon men quite in the Shade but do business just the same.

Our company, *Proprietors of the Invention for Raising Water by Fire*, is known to many. Our Secretary and Treasurer is John Meres, a good friend of the late Captain Savery and our Committee includes Edward Wallin, a Baptist and dear friend, who is the pastor at Maze Pond in Southwark and who assures me that in London the coffee houses of Birchin Lane are where business is well-oiled and sealed. Wallin hath me in a waistcoat and insists my breeches hath buttons and buckles at the knee, the better he says for Real Money to take a Devon Ironmonger seriously. I am accustomed to the rosted spice of coffee beans - amidst gossip and whispers, I saw that men of all shapes of Creation obtained much relaxation in the exchange of ideas and opinions and were free to talk business while they were at their leisure. It hath passed that I hath done business where I thought a God-fearing man would never be, among the smell of muck-worm and smoke as strong as a Dutch barge. I took up a puffing engine myself and became a Sifter of Paper until I looked like a printer's apprentice from the ink black on my fingers.

There are daily as many pamphlets as a man can read and soil his Mind as well as his fingers with as much gossip as business!

It is not well with me that such Business hath Impressed my Mind with all manner of Anomaly's none of which are my Soul's chiefest concern. I rejoice to hear my family is in good Health, and I will insist dear Wallin conducts business as needs be. I cannot wait to be back in God's clean air with God's people in Dartmouth.

I am wary of gathering Treasure for my own Vanity but look always to see it put to God's purpose. I hath said that I will not be the fool, the one our Saviour described (Luke 12) [1] who, when his Soul comes to be required of him, shall be found only to hath been laying up Treasure to himself and is not rich towards God.

Our Engine, which is by rights, a child of many Fathers, hath now taken my name and is called the Newcomen Engine. Hannah is very dour as to the praises heaped upon us. She says Steam Engine hath been my Wife for many years and few would know that her husband was a stranger to her and prefers the smell of Smoke and Oil. That she hath young Elias, Thomas and little Hannah running around her feet seems no contradiction to her Abandonment. She hath a knack of bringing a smile to my face even as she teazes me for some new Defect she hath found in my Character, and I will soon be able to buy her more of what she desires, nay, deserves, now business comes to our door.

I do preach to our Baptist Brethren that Christ is the rock upon which our Church is built but dear Tom Savery, God rest his soul, hath been our foundation stone. An honest man, most generous with his friendship, met his God several years ago and reaped the rewards of Heaven surely due to him.

And our rewards surely grow. Our engines work well enough for collieries to pay a yearly Stipend. One colliery hath agreed to £200 per year and half their Net Profits for our services in attendance upon their machine. We are well able to release ourselves from the bondage of debt even as the Executors of Mistress Dottin, who gather lyke the Pharisees of old, demand repayment of her particular debt and will not accept my dear wife's ledger records that we hath indeed done so!

Wallin claims that we are well placed by Savery's Patent to say, whilst no other company can devise such steam engines, it is also true that no other can manage their temperaments. He hath an eye for business and is far ahead of Calley and I, but I must report Calley hath bought a smile to those who know us well. With the straightest faces, he says he will never jest concerning what we call the Near Catastrophe at Coneygree when our first engine was winde-logged. He hadst me claim it hath left an indelible mark upon our memory and hath caused us to draw a *Table of Temperament* for our machines. The making of this Table hath filled in our time in London and given us some amusement not bent to sully our Souls. We supposed there to be a connection between the local atmosphere and the happy running of our 'atmospheric' engines. Mistress Wallin, in particular, says she is greatly informed about matters she never thought possible on this Earth!

At Dartmouth, not a man, woman or child is long from the smooth waters of Old Man Dart where we can wash our faces and rest on grassy banks and listen to cows lowing. Thus, we are Sanguine in type. Given such active passion and sociable outlook, we looked to our engines to be a faithful copy of their makers but, alas, that is not the case. We find our machines set in the tin and copper mines of Cornwall to be of Choleric Temperament, being given to short-tempered spurts of activity, irregular and irritable. Whereas the Machines installed at coal mines in the North near Coalbrook Dale, Warwickshire and in the district of Newcastle upon Tyne are Melancholic in Temperament, being quietly settled in their business and not given to rash outbreaks of blocked valves or split cylinders. The closest to being Phlegmatic, that is, relaxed and peaceful, are the lead mines in *Flintshire*, Wales, and *Derbyshire* near the green forests of Nottingham where ogres and goblins no longer live.

I do confess there hath been great humour in the reckoning of our *Table* and our men nod and mutter over it. My wife hath written, no matter Mistress Wallin's 'swoon,' she trusts her newly-minted Engine-makers are Sanguine enough not to publish it for there are many more debts to be paid before we are 'Lords in our Manor.'

Our agents tell Proprietors that from a few rudimentary Principles we hath delivered an exceptional Pump but we will not claim our engines are without blemish. It is clear to any Mine Captain who can do his numbers, such atmospheric engines are in want of more efficiency. Our engines grow more robust with better casting but that is their limitation, they burn great quantities of coal day and night. Yet it is a Blessing that these Isles hath strong coal and ore in abundance and it is upon these natural wonders that our Prosperity will grow. Today it is obvious to any man who cares to look that Steam Power hath brought prosperity where there was struggle. Our engine hath shown itself to be five times the pumps on waterwheels and it matters not the lie of the land! I hath heard that Collierys boast that output hath increased to levels never before thought possible, and swear they hath doubled their output. It is easily reckoned that mines can go deeper and miners work safely free from the chill of rising water and all the horror it can bring. It is our engines that keep these mines dry so we hath become, far more than Savery dreamed, a miner's true Friend.

Calley declares our greatest Advance was the injection of water via a water spray to aid the cooling time. He allows that a leak happened by accident but says we were Sanguine enough to know the usefulness of its effect! For myself, I marvel at the cleverness of our engine men who asked if we would consider the application of a rod or pulley to move the values. Our first engines needed Plug Men to open and close the valves, or steam chests, and put an onerous burden on them for timing. They thought we could make a 'working plug' of a piece of timber slit vertically to engage the handles of the valves.

It was late in the day they came to me and were full of apology as to their forwardness. I could hath clasped their hands and said a prayer for their sagacity, instead I told them to wait and returned with a rough drawing of what they hadst described. We were tired and dirty men and I asked them if I could say a prayer of thanks and they said why not, Brother Thomas, do you think the Lord would allow a celebration and took me to The George in Borough Street on my way to Wallin's house and sang me many songs.

A Plug Tree now runs alongside the cylinder enabling the valves to be opened and closed by the movement of the beam and not by the movement of a man's hand. Thus, we enabled the engine to open and close its own valves and I feared not, rather I marvelled and rejoiced at the cleverness of men.

God hath given us the forces of Gravity and Atmosphere, but that we could invent a machine truly self-acting still astonishes me yet it should not, for God hath given us Dreams which advance our Knowledge. That He hath preserved for Himself all manner of Connections entirely unpredictable will keep our Vanity in check.

I must record also, we hath become users of Iron far more than Brass, as Brass is costly and not able to bear the size we think necessary for larger cylinders. Any man who deals in steam must be wedded to iron for we are bound together and must bend it to our Will. The first man to bore a cylinder true to its specifications and free us of the leather seals currently in use will be the true master of our engines. It is a long way to the forges of Coalbrook Dale, and it is many weeks from the comforts of home, but that is where the future lies, in the hands of Ironmasters lyke the Quaker Darby and the iron men in the Severn valley, whose furnaces, I hear, now burn through God's night and light the skies for miles until dawn.

***

REALISATIONS

PART 2

Steam Team 2

John Smeaton 1724 - 1792

Matthew Boulton 1728 - 1809

Joseph Black 1728 - 1799

Erasmus Darwin 1731 - 1802

James Watt 1736 - 1819

1730

By 1730 the great scientific minds of the 17th century were dead with few but Isaac Newton living beyond seventy years. The Enlightenment now simmered; new knowledge lay potent and ready to flower into nothing less than a revolution. The population of England and Scotland already exceeded 10 million as large-scale agriculture turned farmers into city dwellers; fed on the dreams of a trading nation, they flocked to villages and made them cities. News spread, for those who noted such things, that the last native roe deer in England was hunted dead at Hexham, Northumberland. Meanwhile in London, the great scourge of gin addiction was fed by over 7,000 breweries offering relief from the pains of hunger and drudgery, and in excess, a gut-wrenching death.

Across the countryside of Cornwall and Devon, north to the Midlands and scattered through the Continent, Thomas Newcomen's giant engines towered over mines, devoured coal and kept flooding to a tolerable level. They were part of the mining world, anchored deep in the pits, their pistons labouring on the fragile coalition of vacuum and gravity. The smoke from their chimneys and the dull repetitive thud of their pumps became a part of a new landscape, tolerated because they generally worked and made their owners rich.

Thomas Newcomen lived to see his engines become the success he wanted, but he died in the house of his friend Edward Wallin, relatively poor and uncelebrated in 1729. He is buried at Bunhill Fields on the outskirts of London, his grave site unknown.

Robert Boyle did not live to see the new century, dying of a stroke in 1691 a week after his beloved sister Katherine, Viscountess of Ranelagh. They had lived together and shared their intellectual pursuits for over twenty years.

They are both buried in St Martins-in-the-Fields. Christiaan Huygens, the Dutch scientist, died in the house his father left him on the outskirts of the world he despaired of in 1695. Denis Papin did not survive the defeat of his life's work, dying in obscurity in London about 1712 or 1713, history being unable to know for certain such was the depth of his destitution. Gottfried Leibniz lived until 1716 and was buried in Newstrader Church graveyard in Hanover but his grave site is unknown.

But the next wave of inventors was forming, men who would carry their dreams into the world we recognise today. Robert Dick Jnr was matriculating at nineteen from the influential university of Glasgow. He would become an honoured professor and recognise in James Watt a rare ingenuity. James Brindley, the home-schooled mill worker who became a celebrated canal builder, was an adolescent of fourteen years, John Roebuck, the chemist and innovator was twelve. John Smeaton the renowned engineer a child of six. Matthew Boulton England's great entrepreneur, Joseph Black, Glasgow University's great chemist and John Wilkinson the Iron-maker, were infants of two years. The remarkable potter Josiah Wedgwood was newly born. He would mass produce pottery and make English ceramics internationally famous.

The story of steam begins again in 1755 with the anxious James Watt at twenty, struggling to finish his training in London with no apprenticeship papers and a small allowance from his father. By then the remaining members of the Lunar Society, Doctors William Small and Erasmus Darwin and the philosophers and chemists Joseph Priestley and James Keir, having survived childhood, were making their way in the world. Like Thomas Newcomen before them, many of these men rose without a university education. James Watt, Matthew Boulton, John Wilkinson, John Smeaton, James Brindley and Josiah Wedgwood were men of talent and enterprise who suited the times and turned it to their advantage. They were proud and ambitious and the world was their oyster.

*

21

'my heart in my throat as the world's horrors opened to me'

James Watt is the son of a respected shipwright. He is 20 years old and has travelled from Greenock, Scotland, to train as an Instrument Maker.

James Watt
London, 1755

I held my breath lest it make noise upon the air and pressed the Tavern's wall hard to my cheek and wished my bones melted into its ridges and prayed I might vanish like smoke in a sudden wind.

It had gone nine hours by the chimes of his clocks when I left Mr. Morgan's workshop and the twilight made shadows in doorways and spread itself to narrow the lanes as if to say a citizen about his business on the public street should take heed, which I had not.

I first heard their boots clattering on the planks of Threadneedle Street like a sudden heavy hail and I knew their purpose before they came into view and turned towards me setting the stones sparking. They carried a sour damp with them that made my stomach turn. I cowered as a ghost invisible in Finch Lane for I had not one brother guildsman to defend me. Holy Mother of God let them pass by, I prayed, and so they did, but I stood with the beat of my heart as a kettledrum in my ears and I did not move through fear of being unsteady on my feet.

This is a city of dung heaps and chamber pots so rich on the nose in summer the trees should wither brown to mimic the air, and for all the necessity of my being here I would never reconcile being Pressed among such crottles of men who relieve themselves in the streets as freely as animals might or fall like dead men blind drunk for tuppence.

My mother, forever in my mind, would think me foolish to work so late but it is the fate of an apprentice without papers seeking to better himself to be easy Prey for burly Pressmen looking to satisfy some desperate Captain. I will not send this on, for my father would worry greatly to learn that his son kept such hours and he would surely admonish me as cavalier with my future to be out after sunset in such a city. Not being men familiar with the charms of London, I doubt my father would ever suspect how many men are brought before the Lord Mayor himself and, being without apprenticeship Papers, have no grounds for appeal against their Duty to the King's Navy. Or worse, are knocked solid by a Press gang and wake sodden in bilge water, in a Man-o-War, with an ocean deep under them!

My mentors are the dearest men to me, even though I suffer at their advice. My father took the advice of Professor Dick, who we know through Professor Muirhead of my mother's family, to send me to London, there being, in all of Scotland, no master to teach me instrument making and keep me safe near home! I did carry a useless Letter for Instrument Maker John Short, but he would have none of me. In London's Worshipful Company of Clock-makers desire cannot speak above principle, and no outsider will be taken on unless he is first apprenticed to a Freeman of the Guild. That John Morgan has taken me on, even though it be in servitude and that for a fee, is a training I must bear.

My hand is still a tremor and has splashed the Brandywine my landlady insisted I drink. Tonight, I came within a hare's whisker of being kidnapped and I should be laughing that Fate has been so kind to spare me from that miserable Doom but I shake and berate myself for being on the streets at such an hour.

Besides there is another shock entirely. I would not have it known in Greenock that I prayed to the Holy Mother with my heart in my throat as the world's horrors opened to me.

*

22

'I would have signed a thousand Papers'

John Morgan is a Freeman of the Worshipful Company of Clock-makers and a Master Instrument Maker.

John Morgan
Finch Lane
London, 1756

My wife would chide me if I had not offered him a decent meal before he departed. Which I did, but had to insist upon him having it, a good soup and a strong Ale for his journey ahead for how many days I knew not, on blasted Roads to northern parts beyond even Glasgow. James is a raw-boned, serious lad whose smile is a rare occasion, as my wife notes, as if it was my fault he arrived begging to be taken on against all proper Conventions of the Guild and has hardly smiled since. He worrys me, she says, he should smile more while he still has good teeth. She clucks about, making me wait for my breakfast, while she considers her next complaint and sings the praises of my assistant.

I do not sing praises but James Watt has proved himself a worthy assistant, being most thorough and skilful. He outshone my apprentice who is slow to master items Watt consumed in a matter of weeks. He asked to be taught all branches of the Business and had no trouble with his Mathematics or his hands to work on slides, rulers, sextants, scales and quadrants Et cetera.

Mrs. Morgan has words aplenty for working him hard in a cold workshop, cold she says her Scottish mother would call a *yowe tremmle*, as if he was a freshly shorn sheep set trembling for his trouble and I was a cruel farmer to leave him so exposed.

Yet I would not tell such a Craftsman dedicated to learning his trade not to use every Hour of the day available to him. It was entirely his nature to arrive with the morning Light and not leave until Dusk. If truth be told he lifted my Spirits mightily for it had been a long while since I had delighted in the skills of my trade executed so precisely before my eyes. I have often lain awake these last months devising what further exercise I could present to him so he may not look at me as sometimes he did, as if I had given him a Child's task to solve.

After he had eaten I stood like a father and brushed crumbs from his jerkin and told him that I would sign a thousand Papers for him if I had them to sign, for he had learn'd as much as I could teach him. He shook my hand most heartily and said it had been an honour for him to be taught by a master such as myself, a man who had made a Telescope for the King of Spain and written so well on the problems of Time and Longitude. That is what he said of me, Bless him.

I did watch him walk down Finch Lane and felt my soul shrink. It did not rest well that I had charged him Twenty Guineas plus his labour for these many months but I knew that if I had gone after him, he would have been too proud to accept any payment for his services. I came inside to the shop and my wife was dapping her eyes and told me, you are a hard man John Morgan. I did not answer for I had none she would not distain, even as I did myself. We should pray, I said, for his safe journey back to his family.

*

23

'It has been a week of great fortune for me'

James Watt
Greenock, 1756

I am blessedly home in Greenock among family and friends, the air fresh with salt and the public streets safe to walk at night. There are friendly faces about and our houses are well-favoured and trim and I am suddenly most fond of them. Yet I am surprised to see beyond the St. Laurence's Bay, the mountains of the highlands rise as if they had been arranged so in my absence! I had not remembered them looming so and I have walked the quay as if I am newly arrived to my childhood. Across the Firth a wilderness lies with high dark hills where, (we were most sternly told) the savage Robert Roy MacGregor fought with broadswords and dirks. After a year in London's streets I would happily wander those dark hills with their hairy clans; it might make me more of a Scotsman who would not blink at the savagery I found in foul cities.

Meanwhile my father worrys me to eat and rest having seen what he calls his Scarecrow Son returned from London. He sends my sister Jean on errands to find me nougat or sugar plums to aid my digestion and sweeten my disposition and fusses about like an old hen more than he ever did. As for my progress in Business there are no fertile fields to plough my trade. Times are harder than I thought and men who should know better seem against me. My father reports that the Corporation of Hammersmiths in Scotland have refused my petition as I have served no years of apprenticeship. The Guilds have a high standing and I will have no work without their permission.

My hopes rest in finding a partner who will fund my Endeavours, so I may be able to ply my trade and be my own Master. I have long desired a shop in the Saltmarket at the end of High Street, which is busy and much favoured. There would be my name on its door and a door bell to charm the entrance of those wishing my services! If I were to sing my dreams it would be to the happy little peals of a shopkeeper's bell. I would obtain a most comfortable chair for my customers for my mother once told me people are the happiest when they are comfortable. My hopes rest with Professor Dick. My father is still on great terms with him, and since he advised for London I am hopeful that he will want to invest in the much-improved craftsman who has returned.

*

Glasgow, 1756

I came down to Glasgow with rain buffeting all about. I felt no more than a school boy with my lunch in a cloth from my sister and coins provided by my father clinking in my pocket. It has taken but a few hours and I was counting my Blessings and I am still stupefied and cannot find a kerchief dry enough about me for happy tears. My father said Robert Dick would welcome me so I called upon him. He rose from his studies and came towards me and held out his hand in a most friendly manner and said, you are just the man I needed to see, Jamie. Just the man.

I said I was at his service in all things and gave the deepest bow my aching body could devise. He looked most pleased and Lo, he offered me a most prestigious task.

He has called on me to repair Astronomical Instruments much damaged from Jamaica. To be paid to do what I know I can do tolerably well is a great joy to me. To be given the honour of working on Sir Alexander Macfarlane's own instruments [1] for studying the stars is an even greater joy. Given that they have endured a long journey by sea I suspect they will be much salt affected. Brass, being a copper and zinc alloy can corrode, as I have often seen in my father's shipyards.

*

By way of Robert Dick, I have this day met Professor Anderson who is also well disposed towards me, having heard good reports of my character from his younger brother Andrew, known to me from our school days. Andrew assured me he spoke well of me, though he laughingly threatened to tell his brother the 'Truth' if I did not buy him several ales as soon as I had set up my 'Premises' He warned me to be wary of his brother who he declared was a rosy faced fellow up for any madness with 'experiments' but a fierce dog underneath! I am happily his obedient servant no matter what sort of bark he might have since the Honourable Professor arranged I be given lodgings and a workroom in the grounds of the University. He was jolly enough when he announced the University, being Master of its own precinct, could employ me without the censure of the Guild.

It is well beyond my expectations that I would find work in the proximity of great Minds and great Opportunity. I dare hope I may procure a position as Instrument Maker to the University for if I am ever to have such a position it will be at the call of Philosophers who care more for a man's crafting than his station.

*

24

'a machine powered by steam can be used to run carriages'

John Robison is the son of a Glasgow merchant and studying Natural Philosophy. He is 18 years old and writes from Glasgow University.

John Robison
Glasgow, 1757

There is nothing that has gained my interest so much as power. I think on it constantly and my visions of great advancement follow me into my dreams. In case, dear reader, you might think this a pompous vaulting without the wit to know it, I do not mean the power to move the minds of men or to conquer nations.

I mean a simple power easily measured. I seek a way to move an object from one place to another by machine, a machine bearing the weight and distance necessary for our industry and prosperity. There are some who have laughed at my notions and protested that there is no more to learn since the wind has driven our ships and water our mills for generations back to Adam, and both man and beast have laboured to fill the rest being the natural energy God has given us.

They want nothing to do with change to the established order of our industry but they do not rub well with me. The University is crowded with men of new disposition. They will not be cramped and declare the greater good is to be found by constant enquiry, best done by a mind not sullied by emotion or superstition.

There is great sport to be had in the Department for those who throw ideas in the air and are able to make them land, and I am keen to play.

I am convinced, dear reader, a machine powered by steam and suitably assembled can be used to run carriages and by ingenious rack and pulleys move weights beyond our present methods. If I am a man of courage I would confess that I have seen such a carriage, as it were, in my mind's eye, and it is my fancy to work backwards from this vision and meld the facts to suit such a creation. This is a time, more than any since, for such adventures.

Moreover, I have become acquainted with Mr Watt who is a fine chatty fellow ready to consider such adventures. I plan to entice him to dwell upon the possibilities of steam power. His understanding of such Mechanics and the Equations to construct them is far superior to mine, and indeed, I believe him superior to any man known to me. He has been made Instrument Maker to the University but is the centre of wider Matters. His workshop is frequented by men of great intellect who show no concern for his lack of a gown, rather they hang on his words and watch his steady hands rub life back into precious instruments so Valuable to their Pursuits.

That I am astonished at my Fortune to be considered part of Watt's circle is for all to see. I wonder why Fate has allowed my association and if it were not a pagan habit, I would pray my Gratitude to whatever Muse has smiled upon me.

*

25

'I am the happiest man alive'

James Watt
Instrument Maker to Glasgow University
Glasgow University, 1760

I am the happiest man alive. That is the most certain thing. My mother once told me that a skipping boy is a happy boy and I am much inclined to prove her wisdom and skip to my rooms at the University but for fear it would set my colleagues to much laughter. They may call me *in love* and ask about my wife but I would assure them that a wife is next upon my list.

A happy man has a list and I have a list -

> *Between the acres of the rye,*
> *With a hey, with a ho,*
> *with a hey nonny no.*

My brain is so full of air I may fly away on a sudden wind if I cannot fasten myself forthwith. The shop with partner John Craig goes well tho' he says it will take time for Glasgow to know we offer not only mathematical instruments, but can satisfy our customers with mended toys and the resurrection of their wind-blown bagpipes, rusty flutes and other bruised and battered instruments needing our care. I am be-twixt and between the shop and the University, until my head spins with a thousand notions.

My father is ever anxious of my welfare and I send nothing but good reports. That I have a workshop where I can explore the properties of what interests me is a wondrous blessing for I do not know what will excite me next but I am assured it will arrive Post Haste. I am about, of all things, steam power. It was the scoundrel Robison who dangled the idea, declaring it to be as warm to my appetite as smoked salmon is to a Bishop.

In '57 he was bold enough to publish his opinions in the *Universal Magazine* (1) where he argued that steam could run a carriage. It struck me as courageous to write so in front of his peers and betters. I wondered from what part of him came such a prophecy.

I recall he looked most pleased when I quizzed him.

I knew you would be intrigued, he said.

How so, I asked, fearing I might be a fish to his hook.

Because you have an ingenious mind. Besides, I believe it is a revolutionary idea which will prove to be true.

Then he clapped his hands with excitement and made me laugh at his joy.

If you are willing, he said, we can set ourselves upon such a task and make a name for ourselves.

He then spread his arms wide and spoke most solemnly.

Look around you, Mister Watt, Sir, - Glasgow is full of men who have some guts in their brains. Why shouldn't we join them?

He then let out another laugh and had me laughing with him.

We should sign our pact with something stronger than Tea, don't you agree? His voice echoed in a mighty fashion round my little room till I feared it might spill out into the University and all might know our fantastical plans.

He rose and suggested we could continue our 'collaboration' at the Old College which we did. It was well night when we rose for home and Robison, in a fine humour, pointed out the high white moon which would light our way. He bade me a fond goodnight, and declaring me to have the far greater intelligence, left me 'to create some figures.'

I walked home smiling to myself and finding I was, as Robison wanted, well intrigued.

*

True to his word Robison was knocking on my door well before noon, clean-shaven and his hair back like a sailor saying he was before me at his earliest convenience to make us men of fortune and fame. I thought he might have a touch of barrel fever from our collaborations of the night before but he assured me he was sane, nay, inspired, by the prospects of our new adventure.

I was not convinced of any one thing he said but his zeal radiated an excitement of possibilities so fantastical they were, if I am honest enough to say, more enticing than anything I had ever contemplated. We set upon our nonsense like men possessed but as much as I could devise, in secret. We indulged ourselves in experiments with cranky cylinders and rack motions fixed to wheels that proved to be successful only in the complete Demonstration of our Ignorance in all matters of steam and motion.

Soon after Robison went to sea, sailing with the Royal Navy no less, in the employ of Admiral Knowles as a tutor to his son. We hear he is with General Wolfe's expedition, gone to fly the Union flag against the French. He writes he will have many fine tales to tell, but I am left with twisted scraps of iron and unsolved failures to plague my mind. I am sure he thought to let the matter rest but now it is upon me I cannot leave it be. At the end of a day it comes floating into my consideration unbidden by any concrete plan known to me and settles itself in my mind. John spoke of a machine powered by steam with such conviction and told me he had seen such in his mind's eye and when he spoke I believed him. I have inquired and found steam power works in the mines and that seems, to learned men, to be the end of it. Now John has left I might as well be Crusoe alone on a Steam Island for there are no champions of its power in Glasgow. There is no one who would know one end of a steam engine from another and I have turned to reading old papers when I can find the time.

The work of Dr. Papin and Captain Savery have left paths for any Wanderer to follow if they have appetite enough. It may be to no final purpose but I have a copy of Jacob Leupold's *Theatrum Machinarum Generale*, (2) written in his native tongue and far less daunting than Latin. I wonder what my dear mother would say if she knew the boy she forced to his lessons and daily complained at grammar school, was now learning German and discussing Leupold's 'Krafft, ohre Kunst, ist hier umsunst'(3) with whoever may listen? Probably what my father has recently written:

I implore you to eat well and keep warm for great men are unlikely to converse within the miasma of a sickly colleague.

But in truth I am at the mercy of fantastical possibilities John saw as merely a herd of late cows waiting to be milked by Glasgow's Instrument Maker! My mind is captured and I cannot walk the Green and take the air as I used to, for a thousand thoughts tumble round my skull and I am home more restless than when I began.

It is a boon to be accepted into the university clubs, in particular, John Anderson's club where the discourse follows paths never before open to me, but they offer me no rest for they offer me no excuse. They will brook no limitations on the minds of men and have me tingling in anticipation of wondrous Ideas. Such men as John Millar and Adam Smith stare Ignorance in the face and do not turn away, rather they declare it a moral challenge to be overcome.

John Millar has me laughing one moment and exalting the next. He says that mankind moves naturally from Ignorance to Knowledge and from rude to civilised manners. I am thankful he does not visit my workshop when I confront yet another failed experiment and do not know if I am monkey, man or beast! Millar has seen through this particular of mankind for marooned in ignorance, my manners resemble a most uncivilised state.

To prove this no exaggeration, I have a small steam machine on my workbench. It is a Newcomen Model given to me by John Anderson who was unhappy with the work done by Jonathan Sisson in London to repair it. Anderson inquired whether I could make it work more efficiently. It now sits in front of me and daily offers an affront to my sensibilities for it is a caw-handed thing which offends my understanding. After many adaptations to the boiler, the fire and the cylinder, I have not been able to make it work in any way that satisfies me. It is clear I am confounded by the behaviour of heat and steam and have no mathematical theories to guide me.

By chance I have procured a Papin's *Digester* to study. It is an odd contraption which at first amused my colleagues, for it seemed born of another world being of considerable age and quaintness. It is a simple round pot, but when full of steam it can hold and release an enormous force by way of a small valve. It is the ingenuity of the valve and the pressure it controlled which most interested me. John is a fine friend to leave me prisoner in his 'steam carriage' but I confess, I see a force here no man has truly harnessed. It has made me a man of day-dreams for there is no questioning its power which I have already measured. By way of a syringe with a small piston inside I made a little piston attached to a rod. It was easy enough to fashion a steam cock to control the entrance of steam into the syringe. Then I watched the force of steam push the rod to lift 15 lbs. of weight! It is one thing to read about a force, it is another thing entirely to see it in action.

Now I see its force I am even more startled by Leupold's high-pressure non-condensing steam engine, a design far bolder than anything I have seen in my studies. Leupold confidently calls such a steam engine not only possible, but preferable in his *General Theory of Machines*.

Leupold had the support of the great mathematician Gottleib Leibniz but I shall have to do with my own poor calculations and they make me wary. I could well imagine Papin's small pot bursting its seams and creating mayhem to its surroundings and since I could ill afford such a Calamity to injure myself or others I have put thoughts of such strong steam aside.

*

26

'a Wake without a body.'

James Watt
Greenock, 1762

My father had wanted me home *post haste*, and so I am.

My dear sister will not keep still and is forever cooking as if we are Inn-keepers expecting a jolly crowd on a late coach. The truth is a very poor cousin for my father's business friends, Jock's school mates and the few relatives such as the Muirheads and the Millers and of course dear Robert Dick and Joseph Black are all welcomed to a Wake without a body for my poor brother has drowned.

Of the two Watt boys, it was the thin sickly Jamie that caused my mother to worry and not the rude health of his younger brother, John, who was Jock to us. Now at twenty-three, he is lost to us, drowned in the Atlantic on his way to America, and James, myself, stands strong as I am able, to make the family name.

My mother's family are here, Professor George Muirhead and Thomas Muirhead and I sense they are as unsettled as myself for father has been unable to remove my mother's favourite trinkets, sketches and such, and the spirit of Agnes Watt, nee Muirhead, is everywhere.

My father has done his best, as he did for my mother's wake. He has ordered a bell be rung through the town, bought whisky, stopped his clocks when the news arrived and covered them with a black ribbon. He also covered the mirrors as he had done for my mother and I wondered if he knew the difference in the names, if the grief was the same, nay, compounded. I may go about as the eldest child but I am a lonely promontory, for there were three before me whom my mother carried, all for no living proof of her loving heart. Now there is only my sister and myself to comfort my father.

My sister has bent herself to endless activity and anxiety to do everything the Proper Way. She poured our milk onto the ground, and out of respect for the fishermen of Greenock, she says, many of whom have known Jock since he was a baby. The butter and onions were also thrown out as their families do.(1) Jock's clothes and keepsakes were folded away somewhere and I did not ask after them.

After the Wake, there would be no body to take to the Kirkyard and I was glad of that. There would be no heavy wooden plate to put on the body with its portions of earth and salt, which would only remind us that Jock lies at the bottom of the sea with salty water around him many fathoms deep.

I have spoken to my sister in the last few days more than I have in years and it caused me to wonder what sort of brother I had become. Jean is a shy lass of some twenty-two years, and known in Greenock as a home body. She told me that father was saying he had seen lights floating over the sea and asked what she should do.

Nothing, I suggested. He believes it was a sign which foretold a drowning.

Do you such signs are true, she asked me quietly.

No, I do not. But do not argue against him for fear of adding to his anguish. Remember he built that ship and will be challenging everything and blaming himself.

And what should I do about the bees?

I knew what she referred to. Some people are honouring bees as charmed souls because of the power of their honey to heal. They confide in the bees, share their loss and drape the hives in black ribbons.

I asked for her thoughts on this custom.

I think it is a lovely custom and I would find it a comfort, she replied.

Then that is something you must do, I said in return, wishing she would take her life into the living of it now she was of age.

As if she read my thoughts she turned to me and said, Jamie, dear brother, you are so very clever we all look up to you, even as mother did.

My face burned apace but it was shame at my neglect of her rather than pleasure in her love.

I pressed her hand and told her that our parents were very clever and that we were Blessed to be their children. She smiled her shy smile and went about the business of seeing to our guests and directing father's new housekeeper who had never seen so many people in one room at the same time.

My father asked me if I could speak the eulogy for my brother as he was sorely pressed. I listened to Jock's school friends as I went about and was able to gather some threadbare words. I have a memory of Jock and I standing beside father when a crane of his devising swung into action over Greenock dock. High above our heads the sweet aroma of tobacco bales moved swiftly into the black hull of some monstrous barge we were too fearful to regard. We hardly noticed as children that all about us were the dry-docks, work sheds and jettys of the Chandler James Henry Watt. For the rest, I could say very little for being so sickly as a child I was kept away from my siblings, schooled at home and cloistered among my father's thousand tools and toys to interest me. As for Jock, he walked free in the world and must have heard the call of the sea across the Firth. He was a talented surveyor and had done much work on the Clyde, being commended by John Smeaton for his skills.

Being taken so young and unfilled, his wake was a sombre affair where no one expected celebration. One of Jock's friends played a sad lament on his pipes and brought tears to many eyes. My cousin Peggy Miller gave me a look of such sad warmth I was surprised at the comfort it gave me. That night I watched the moon's light skip across St. Laurence's Bay as I had done many times as a boy and swore I would name my firstborn son after my brother, John Watt.

*

27

'I cannot believe the luck of the world'

John Robison
Glasgow University, 1763

I cannot believe the luck of the world and all its treasurers.

I am home with adventures under my belt and lessons learned. Nothing has changed for any man with curiosity and strong blood in his veins can make something of himself, and for the greater good. The university has offered me a place for it is said they have trusted in the law of probabilities in my case, assuming my mathematical skills must have improved with all the navigation done on a rolling decks and surveying through what they imagine as devil-infested jungles!

It is true that I have met many great men, including the late General Wolfe but above all, it has been my great privilege to represent the Board of Longitude and to work with William Harrison [1] on measuring the astonishing accuracy of his father's Sea Clock when it arrived in Jamaica in '62. We measured it with great care and showed it out by a mighty five seconds after 81 days at sea!

This should change everything, I remarked to William, the Board should bow to your father's genius for there is no other clock-maker alive who could build such a magnificent time-keeper. I am honoured to say I was the man to measure such brilliance! I remember his uncertain smile and his remark that the Board was pig-headed when dealing with the son of a carpenter. I took that in for I still had faith in my friend the son of a ship-builder. As for Glasgow, the lads at the University are much the same, noisy and full of energy towards any madness which might interest their professors. They smell as though they had run from Marathon and swish around in their scarlet gowns and pride themselves in conversing in the worst Latin their professors can imagine.

Mr Watt seems taller and lads hang off his every word for he has an unfailing interest in exploring whatever notion they may bring to him. He was most pleased to see me and called me at an honourable member of his steam company and suggested I take a keek at his next demonstration.

He laughed at the look on my face, which if it reflected my sentiment, would be astonished admiration.

It is your fault entirely, Robison, he said, that I am engaged in this adventure so you are morally compelled to stay and marvel at what you may learn!

I gave a slight bow and laughed at myself, at Watt, at the boys gathering around us, for I felt more alive than ever I did daring death on the St Lawrence River.

You will see, Watt told me, one of the lads has procured me a battered brass kettle which I fear he has taken from under the nose of his mother. Since steam boils at a lower temperature in a vacuum I need this kettle to help us see which is which.

Which is which?

I had to ask, for I could not help but follow Watt's thread, knowing it would lead us out of any labyrinth he may enter.

What will happen first, Watt replied. Will the cold water condense the steam or will the steam make the cold water, boil?

Will you put a guinea on it, Mr Watt? One of the lads called out.

I will put a guinea on it, Mr Sadleir. But I will ask you to tell me what outcome to put it on.

In that case, Mr Watt sir, we'd better wait until the experiment shows nature's law, the lively Mr Sadleir replied and immediately suffered a volley of cat-calls from his fellows.

Watt cleared his bench and regarded the brass kettle as if he was about to interrogate it as to its properties, and more, its inclinations.

It is a common object, he said, his voice rolling clear out his door and across the flagstones towards High Street as if to entertain any fashionable ladies strolling by.

Let us now see what we can determine with Mistress Kettle, he offered and I half expected him to wave his hands as a Fair magician might.

He placed a bent glass tube in the kettle's spout and connected it to a flask of cold water. As the kettle heated, steam moved clearly down the tube and of itself heated the flask water to boiling point. Watt marked the flask and noted that the steam had raised the water by one sixth.

We can conclude, said Watt, that steam can raise six times its own volume of water to boiling point.

A few of Dr Black's students had heard of the 'stolen' kettle and were drawn to watch. They did not seem surprised that steam appeared to hold more heat than water and gathered round most gleefully to explain to Watt Dr Black's theories of latent heat.

Watt grimaced and pulled me aside.

I should pay more attention to what is taught right under my nose, he said.

Professor Black will be happy to confer with his Instrument Maker, I replied, since you built him an organ, much to everyone's amazement that such a tone-deaf fellow as you could do so!

I do hope, Watt said, his memories are sweeter than mine. My adventures with the Mason's organ required even more study and was a torment of my own making, he confessed.

It was shortly after Watt told me Black had been most generous with his time and curious himself regarding Watt's travails with the Newcomen model. I envy Joseph Black, not only for his intellect but for his quiet poise which reduces all manner of graceless boys to silence. He is the Lord of our Academy, unfailingly neat in his person and appears to be on his way to a club of high intellect when he is most likely on his way to meet with Brewers! Merchants are forever at his door asking for practical understandings, particularly the whisky distillers, and it is for them he ponders the various properties of heat.

Black confirmed, as I knew he would, that as water continues to boil its temperature does not rise but heat continues to build in the steam it produces. This is what he has called 'latent heat.' It follows that there is more heat in the steam than in the water itself. If the steam cools the heat is lost. Watt confided in me that since Newcomen's Model relies on a vacuum created by cooling the steam to move a piston, heat will always be lost as the cylinder cools. He believed that the principles of Newcomen's Atmospheric Engine were at odds with each other and would not be overcome by his tinkering at the edges.

I watched as Watt put the Model to one side; placing it behind some long jars so it would not prey on his mind and goad him, as he said, into wasting long hours in a Meditation for a Solution entirely beyond him.

*

28

'as far as the Herd's House when the idea came into my mind'

James Watt
Glasgow University, 1765

This week Providence led me to walk the Green with the breeze fresh to my face. It was the Sabbath and I took exercise to enjoy the quiet of God's Creation and muse at random to ease my mind. I might have been the only man for miles so absorbed was I in nothing in particular when a calculation came to me and I was brought to a state of Rapture. (1) Not caring who was my witness I put my head back and laughed at the sky like a madman.

As I passed the Herd's Hut I considered lying upon the soft grass to stare at clouds like a boy free in the world. In the pausing it came to me, swift, like a restless fish that had jumped into my hands. By the time I walked to the Golf-house I knew that if the laws of Nature held true, my idea would work and I could unleash the power of steam.

I had been musing on Newcomen's model (it often beat like a pulse in my brain) when the whole cumbersome process of his engine was suddenly revealed. I saw its very bones and muscle. I saw what it lacked and how I would solve such an obstacle. The time waiting for the cylinder to cool and condense the steam caused the great loss of heat. To replace the heat more coal in the Firebox was needed, all to great cost.

Firstly, I asked, was it not true that steam, being elastic would quickly fill a vacuum? And secondly, if held by a specially made vessel, could it then condense separately from the cylinder? The answers appeared self-evident and I want to laugh again as I recall my syllogism. My plan would allow the engine to work constantly since there would be no waiting to cool the piston and the cylinder walls as Newcomen had managed with his internal spray. The process needed a separate vessel which I have called a Condenser since that is its function. By the time I was back in my rooms at the university I swore that if I could make such an engine I would advance the cause of the university and prosper in the process. Anderson would have the university build its name on its Advancements for Industry and I stood trembling among my bits and bods that I might make it so.

My calculations have proved true. This week I have assembled a working steam engine in the back court at King Street entirely more efficient than any hitherto known. I am loath to shout about it for it makes me breathless to consider I have bettered Newcomen! When Anderson recommended I be given a model repaired poorly by Jonathan Sisson, he asked me if I might do better. He gave me a look, not at all quizzical but direct. It was afternoon and he had come straight from his lectures to drink tea. I knew he meant that I might do better than Newcomen himself rather than simply correct Sisson's caw-handedness.

To write of such things sends a quiver across my page. This rapture has stayed with me all week and Peggy has shared my joy. Society calls her Mistress Watt but she will always be darling Peggy Miller to me. We are as cosy as two doves in our nest at Delftfield Street and it is the greatest joy to me to have a wife of such warm intellect. I had not thought myself to be so happy as to be married for it is the lightest burden imaginable, even though it makes me feel heavier, that is, more substantial. Peggy is a strong spirit with a sharp eye who, even after regarding life's flaws, prefers to look upon its small perfections, of which I appear to be one!

I will keep this model from prying eyes until I am ready to show the world. Since the good Lord has seen fit to inspire my calculations, and I am now a husband and father to baby John, I pray He will also provide fruit from their harvest.

*

29

*'that all warm-blooded animals have
arisen from one living filament'*

Erasmus Darwin is a doctor, natural philosopher and inventor. He was educated in medicine at the University of Edinburgh and by 29 was accepted as a Fellow of the Royal Society for his papers on electricity and medicine. He is 36 and robust of mind and body.

Erasmus Darwin
Lichfield, 1767

I am under instructions to wash my hands and refresh my face with lavender water before I proceed to my study for my good wife insists she married a gentleman, not a jack-of-all trades or some wild-man from Africa. It is a simple enough procedure, the washing, the linen left for my use and the sweet tang of lavender, but it works wonders on my scattered thoughts that my face is smoothed like a river pebble from the irritations of the day. It does not take me long to settle and read with new eyes the thoughts that I may have left half-baked upon my pages.

There is a Duty among my noble Lunartick friends not to waver in the pursuit of Knowledge and to follow only where Reason herself leads, so I have become a veritable badger with my nightly digging into the great questions before us. This is a comfort to Mrs Darwin, for she reminds me that badgers will eat almost anything.

I am in from my garden where I had set orchids among the herbs, more as a decorative Whim than any sensible plan, and remarked to Mrs Darwin that plants settled best in what suited their needs as I hope to shortly myself being in need of an early watering with whatever Nourishment she could provide.

She protested it I would be a lucky man if she bothered with a jug of water but shortly after she set me in my Study with a good jug of family port and the freshest cut of ham sliced fine between her wondrous crisp biscuits. This time of the day suits me with its mellowness, in either the warmth of the house or the faint scuttles from the kitchen, to accompany my thoughts of Supper. It's here, among my scribblings that I can make sense of the shards of this and that, broken as salient headlands, from my reports.

There is a thesis which grows like a curling Ivy through all I survey, seeking to bind itself to my thoughts. It is clear to me, as it would be to any man who takes the time to observe, that living creatures are bound together by three essentials, these being Lust, Hunger and Security and that without the fulfilment of each of these in equal measure life is indeed brutish and short, as the somber Mr Hobbes reminds us. It follows that since the needs and impulses of all living things are the same their source must be the same, and so bound by natural laws since Creation itself. This notion will provide much stimulation for the Lunar men at our Circle meeting. Mrs Darwin tells me I have been blessed with a bold imagination but there are minds more inspired than my own. And it is about one of these minds I now write while our meeting is still fresh in my mind.

By way of my son and the remarkable William Small, who recommended my interest, I met today with the engine-maker James Watt. He was returning to Scotland from London and agreed to stop at Lichfield. I was in great anticipation for I would not fail to follow any recommendation from Dr Small since he has inspired young Erasmus to hang off every word he says. And my son is all the better for his adoration. Small's intelligence flows evenly towards all who speak with him, young or old and I agree with Boulton, we are fortunate he has settled his Practice in Birmingham.

After Watt took his leave I went straight to the herb garden and observed their humble evergreen purpose, both medicinal and gastronomic, then crushed some thyme leaves and held them to my nose and breathed their fragrance hot and penetrating, for I could have fallen on my knees and smelt the rich earth and cried Halleluiah, Halleluiah. This day I have heard the Future speak and all the Glories it will bring for Watt is the man to produce great machines which will move the world!

*

I am busy Doctoring of late for the weather has been so obtuse as to have been called higgledy-piggledy by Mrs Darwin, and left half of Lichfield out of sorts - but my notes regarding Watt will be of great interest to Small and Boulton.

He is, on first impression, an unremarkable man, tall but seemingly without great strength and given to a slight stoop as if he has just this moment risen from some weighty tome. I entertained him in my study and found myself anxious to put him at his ease, for he seemed burdened with concerns he was reluctant to express. Dear Polly was Charm itself and fussed about him insisting he sniff Lavender for his Headache and drink her sweet China Tea for his refreshment. He was awkward at first with his responses for I think he was not fully informed of my interest in steam. He seemed reluctant to discuss such interests and told me he had sold his business in Glasgow and had set up an office in King Street as a Surveyor.

Whether wise or not I could not contain myself and spoke of my great passion. After a few minutes of my explanation of my own idea for a Fiery Chariot and what I thought would be the principles of its operation; my concerns regarding the number of wheels and whether Power should act on one wheel only; my concerns as to the management of the steam Cocks where by the Motion might be accelerated, retarded, destroy'd revived, all instantly and easily; and my concerns regarding the amount of steam pressure which would be required to move such a vehicle, he interrupted me with a great smile across his face.

Dr Darwin, he said, I must be in a dream for I never thought I would hear a man so advanced in his knowledge of steam power.

My dear man, I replied, although we may not yet have a University, you are in Lichfield which is close to Birmingham, and Birmingham, I can promise you, will be the centre of all Advancement. One could say we are Infected by Steam Enginry and tho' I am Doctoring I can find no cure for it.

This time he laughed, and said warmly, dear Doctor, you make my heart sing even if you are a Master of Hyperbole.

Touche, I thought, but I did not mention that our Circle had been alive with steam notions, crude and undigested as they might be, for these five years past.

You have met Dr Small, who has a great mind, but you have not yet met Matthew Boulton who thinks only in ideas of a Universal Dimension, and has the confidence and funds to realise them. You might have heard of his manufactorary at his estate in Birmingham where he produces the most agreeable dinner ware for an elegant table. He would be a great friend to any man who can give him a steam pump to keep the water to his watermills.

Watt seemed amused. Indeed, Dr Darwin, he replied. Water is a constant bother. We either have too much of it in our mines or not enough of it in our rivers.

That is indeed true, I said, and laughed for he was a most interesting fellow. Despite these difficulties I do, most respectfully, recommend you meet with my good friend Boulton.

I would be most happy to meet him at your kind recommendation, Watt said, although I am engaged with John Roebuck on my enterprise.

I fully understand, I said and silence fell upon us.

Watt looked about as if he was about to take his leave. Instead of rising to my feet like a sympathetic host I continued for I wanted to know more from what Small had described as a man of most particular talent.

Boulton and I have often discuss'd what Industry may need and we always find ourselves back at the idea of the turning wheel, I continued.

Rotary motion is a Fancy of many and you are not the first to mention it to me, Watt seemed to sigh.

But think, Watt, of what we could do with rotary motion driven by an independent portable force, I insisted. We could free ourselves of the water mill which ties us to rivers. We could be free of the labour of animals. We could build to our heart's content!

Watt laughed. Dr Darwin, I do not know how I would manage being at the mercy of both you and Mister Boulton.

But Watt, dear sir, I am deadly serious. If steam power can be made to work there is no telling what it may bring!

Watt leant forward. If I can impose upon your confidence I would give you some intelligence which may be to your liking. He almost whispered as if he thought enemies of the Realm were at the door!

I leant forward also, my mind agape. You can rely upon me, Watt, my dear fellow, I give you my assurance.

You will be pleased to know, Watt continued, I have already assembled a working model which can better any known steam power. All I need are the space and the funds to build on the necessary scale and I have a partner supporting my enterprise.

You have an engine that can better any known steam power?

Yes, I do. By a factor of a least two-thirds.

That is a considerable factor, I said, amazed. And your partner, I enquired.

It is Dr John Roebuck, the chymist who has bought into Carron Ironworks near Bowness, Watt replied.

Roebuck was known to me but I did not offer comment instead I asked has he thought to patent his machine. I explained that my own idea of creating a Vehicle on Wheels that can be propelled by steam power seems of such common sense, I was certain not the only one who has considered it.

I could not help myself for I added, perhaps too frankly, that I would not want a fellow Inventor to usurp my ideas.

Watt was momentarily silent.

I would not abide any man who fails to honour my honest work, he said. I can assure you, he continued, as I would never use another man's designs, I will fiercely defend my own.

Indeed, my good fellow, was my reply for he was indeed fierce and I thought he might rise to take his leave, seemingly affronted by my clumsy inferences, but he did not.

He continued, albeit coolly.

As to my Patent, I am not hopeful. This last week I have failed to get a Bill passed by parliament in London for our canal route from the Forth to the Clyde. I have come from inspecting the Duke of Bridgewater's canal.

And what did you make of it, I asked, happy to move to new fields.

It is a great credit to the Duke, and especially to James Brindley, (1) and to his workmen, he replied, being more at ease.

Its design is commendable, especially so at Barton where Brindley has solved his problem by an aqueduct over the River Irwell at 39 feet! I do not know if I would have had the courage to attempt such a thing!

His hands formed an arch as if he were imagining Brindley's victory over the troublesome river and his face glowed with ready animation.

I heard he is a great innovator with his puddling clay to line the bottom of his canals, I continued on the same theme for I had a sudden and unaccustomed desire to impress this sensitive Scotsman who seem'd not to want of any knowledge on a dozen subjects.

He is a self-made man, said Watt, and works for the future. By using a canal, thirty long tons pulled by only one horse will halve the price of coal. Mind though, Watt quizzed, halving the price of coal is a favourite song in these parts!

You are among Canal Men in Birmingham, I told him most warmly. We will support all ideas to Advance our Economy and we will work to halve the cost of all we can. As for steam, it is a whole different matter than the vexing tryals of canal routes. I can assure you Watt, you are among the Friends of Steam, in Dr Small and myself. Can you imagine, I continued, if your engine can speedily pump water from mines and the locks, it will change the cost of everything.

Dr Darwin, I have only gone as far as wondering where there is a foundry that can make a useful cylinder. Again, you have stopped at the right place, I laughed, for Birmingham is surrounded by Furnaces and their masters who boast they make the best Iron in the world.

I do have something that might interest you given you are a gentleman of such varied interests, Watt said rather shyly.

Pray tell, I said, most intrigued as to what he may think of interest to me.

Last year I made what I call a perspective machine, very rudimental, but I have been told quite serviceable, for I fashioned it to be portable.

A portable perspective machine, I inquired.

Yes, my dear friend Joseph Black who has recently taken the Professorship of Medicine and Chymistry at Edinburgh, asked if I would consider such a task. It is on three legs and has an adjustable arm, its own pencil and can be folded completely into itself. Perhaps it could aid your study of flowers and so forth, he suggested.

Such a machine most certainly could, I declared, and began to wonder what else this man sitting in my study would casually introduce. I would be most interested, I said being quite moved, to use such a practical machine.

Watt smiled and for the first time stretched out his legs as if he felt at home. I leant forward with a most serious expression on my face. I do have an important question, I said but did not wait for his response. Are you an Ale man or a Port man, I asked. There was at last a twinkle of warmth in Watt's eye.

I will leave it entirely up to myne good host, he replied.

In that case it will be Port, I declared, and make yourself comfortable for I am now your Innkeeper for the night.

*

30

'counting the shillings slipping from my grasp'

John Roebuck has an MD from the University of Leiden, Holland, but is more a chemist, natural philosopher and inventor than a doctor. He is 50 years old, a Fellow of the Royal Society, and has recently bought shares in a coal mine.

John Roebuck
Bowness, 1768

It is a great grief to my Purse and poor sight to see men sitting idle as a dark swell of water swallows their livelihood and threatens the shareholdings of good partners. We have sought to expand the colliery at Bowness to mine new seams, but all we encounter is sunken rivers seemingly intent on rising to their former glory through our fresh dug shafts.

If a Crystal Gazer had foreseen my present state of exhaustion wrapped in stiff oilskins against the cold and suffering my riding boots to be at the mercy of grit and slush instead of a comfortable day seeing to languishing females, I would not have believed a word. I had no anticipation when I graduated in Medicine I would make my fortune from a Colliery. That this is a constant source of surprise to all is well remembered by my wife, who takes some satisfaction, she assures me, that our prosperity is an even greater astonishment to the matrons of Sheffield society.

I have given the blame to Professor Joseph Black, the beautiful chymist. His lectures beguiled me to the mystery of gasses and solids, such is Chymistry and since Chymistry seeks nothing short of naming the mystery of All Matter, it made my days spent in Medicine a poor cousin by comparison. It was Black who told me of the engine-maker Watt several years past. He did not hold back in his praise of Watt's capabilities with steam. He assured me that Watt's advances in the design of engines place his invention far above our Newcomen engine which daily struggles against our rising water and eats a mountain of coal for its breakfast. He further assured me Watt is well connected for his partner in steam experiments is John Robison who has recently taken his chair as Professor of Chymistry.

When I am up at Bowness standing in the miserable cold and counting the shillings slipping from my grasp, I have only Watt's assurances as a remedy for my misery. He is a gangly fellow given to a perpetual frown and of late his letters do not inspire me with great confidence. He seems to have achieved very little with his experiments beyond eating my money and frustrating my expectations. I have written to say I would want more before I spend more.

*

31

'lunarticks and fellow-schemers'

Erasmus Darwin
Lichfield, 1768

Mrs. Darwin is fond of calling me a Perpetual Dreamer, much to my pleasure. Of late she has declared I have expanded to be a Coach-riding Dreamer as I go from patient to patient, down every road and lane hereabouts until my arse has mapped every pothole and ditch in the parish. Rather than an epithet to my diligence it is a sly observation about my girth which, I declare, is merely evidence of my expanding good health!

She is the sweetest creature ever a man was Blessed with and has a surfeit of Compassion for my Addiction as she calls it for gatherings, meetings and clubs. I do suspect that were she given to writing as much as she is to reading, we would have another Jonathan Swift on our hands to berate us with the follies of mankind, my own leading the list! Yet, if we cannot speak each day, I am a man in a wilderness without a Compass.

Polly cautions me over the huffing and puffing, as she puts it, about the trinkets which can cause us so much Bile, all to be digested at the expense of our families. She reminds me that I can be of an uncertain mood after meeting with my fellow Lunarticks. I may be disappointed that an experiment has failed to achieve the desired results, I replied, but I am always of good cheers for we deal in Possibilities. Actualities may be even better, she responded, deftly skirting my outstretched arm and removing herself from my study where she had set down my lunch.

It is true that, after some communications from my fellow Lunarticks, I am downcast. Small has advised me that all is not well with Watt and he struggles to advance his engine. Neither Small nor Joseph Black have the depth of funds necessary to continue Watt's experiments. Recently John Roebuck has expressed his frustration at Watt's lack of progress although he will not offer him a full partnership. Roebuck has had enormous success at Carron Ironworks and if he were to have the Vision, he would find the Funds to follow.

Small writes that Roebuck's commitment is not on paper and comes in dribs and drabs, and that Watt grows daily more despondent. He cautions that we must keep up our letters of encouragement to him for now his wife is with child again he is driven to consider more gainful employment. Small's faith in Watt never wavers, not that his faith in any of our lunarticks is other than constant, no matter our tryals.

My latest project is to seek a solution for Watt's malaise. Mrs Darwin is in hearty agreement that I should introduce dear Boulton to this Scottish engineer we have all been so taken with. If any man has a credible Vision to change the world, it is Matthew Boulton.

I take my cue from the astronomer John Mitchell[1] who first introduced me to Boulton, a man born and bred to Birmingham. It is a great reflection on Mitchell that he did so given that we are both Cambridge men and Boulton left his schooling at fourteen years. Such an introduction has been most fruitful in its friendship and is witness to our new Inclinations where Experiment and Invention are from kindred spirits and not by circumstance of education.

Mitchell finds time between his studies and his pastoral duties at Thornhill to correspond with me about all manner of things until my mind is as stretched as if he has pull'd my thoughts into one of his black stars. He says he was only a babe in arms when the great Newton died and would give anything to have heard him speak. As Boulton and I have remarked after listening to his theories on earthquakes waving across the earth's crust or his ideas on measuring the mass of the Earth, we feel we have sat at Newton's knee.

We live, I do surely believe, in the most remarkable times ever given to Man, for the spirit of the age is for Knowledge and it thrills me to wake into one of the prettiest towns in Staffordshire, adorned not only by a great cathedral but by minds worthy of its traditions. We are graced with regular visitors including the remarkable Josiah Wedgwood who insists on calling himself a struggling potter come for a cup of tea if there be any about.

Dr Johnson, late of Lichfield and of Dictionary fame, has called us 'a city of philosophers' perhaps one of the few pronouncements upon which I am not in disagreement with the good Doctor. But London, where Johnson now resides, with all its plays, poets and literary masters, is far South of the men who meet to make of theory more than grand ideas to impress editors. I do not know if Dr Johnson is aware of our Birmingham Philosophers who meet monthly on the Full Moon to discuss all manner of topics dedicated to the realization of Progress in all its Forms, but since he is a self-made Legend he may find us of no account.

When Small, freshly returned from Virginia, announced himself, he carried a letter of introduction to Boulton from no less a man than Benjamin Franklin. This caused a great stir amongst us for our name had travelled as far as the American Colonies and men whom we admired, admired us. Small was full of stories from his College in Williamsburg. It made us laugh when he told us Franklin was a Polymath of great renown especially among the colleges for his valiant attempt many years ago to list all the words in the English language to describe Drunkenness. He had reckoned about 228 but none of us wanted to inform Dr Johnson preferring to keep such important knowledge to ourselves.

Given my countenance I am not a man burdened with Vanity but I will say for all to hear that we Birmingham Philosophers are no addled eggs who do little more than turn a page. We are weathered men with tools to build our dreams. I do not doubt for one moment that I am, in provincial Lichfield, at the heart of some great Solar System of thought and meet regularly with minds capable of generating such advancements for mankind as any academy on the face of the Earth. And those minds connected as if by strands of a spider's web, seemingly fragile in the morning's sun, but resilient and perfectly formed to catch whatever ideas fly through our minds. These strands hold together a perfect union of theoretical and practical understanding where experiments into science are as honoured as music and poetry, and delight my soul.

*

The meeting between Boulton and Watt, as I predicted, was amiable in the extreme. When Watt had taken his leave, being determined to return to his family in Glasgow, Boulton turned to me and said, he is an anxious creature, isn't he?

At times, I replied, feeling like a protective father, but he is the most talented man I know.

He is admirable, said Boulton quickly, so as not to offend me. Such a quick and friendly mind, he continued, I was quite taken with him.

He is given to sudden ailments, I conceded, for he worries about his family and has already lost his first born, a son.

Boulton was silent for a time.

I could have bitten my tongue for he has, as I swiftly recalled, lost three daughters and a wife as well and was no stranger to grief. We both know what it is to lose a child, but what man in this world does not?

Who has backed him to date, he finally enquired most business like, with this steam machine of his?

He gave me a measured look all but begging me to see the workings of his mind. And so I could, laid open like a pocket watch before me.

It is no secret that Joseph Black has given him funds from time to time, as has Small, but presently it is John Roebuck who has taken on the lion's share of such funding, I informed him.

Hmmm, mused Boulton. It will not be enough.

How so, I felt compelled to ask.

Roebuck wants his engine for pumping out his mines, nothing more, and he is entirely at the mercy of his banks, Boulton replied. I can offer Watt far more. We should get him to Soho and I will show him what the future can hold!

I could see he had the wind at his back, for his eyes were alight and he threw down my good Amontillado like it was miserable altar wine.

Watt should know I would be happy to back him, aye, back him to the hilt, he said, and waved his hand as if to include all the possessions of my study where currently we sat in the afternoon light.

My dear Boulton, I replied, I will see that Watt is advised of such feeling. Nothing could delight me more that you and he should form a partnership to your mutual advantage.

If his engine proves true I will make him a rich man, Boulton announced, as if it was as certain as the sunrise tomorrow.

And he will make you a famous man, I counted, not to be outdone with the extravagance of Boulton's certainty.

Boulton laughed.

Doctor Darwin, he said. I should be famous just for the knowing of you my friend, for you have more ideas in that head of yours than all us Lunarticks knocked together. In a great good mood we finished the Amontillado and called for more.

*

Wm. Small has arrived and seems intent on staying as long as Mrs. Darwin permits which may well be forever since she has told me many times she adores the ground he walks on as he is such a refined gentleman. I pat my stomach and declare she has far more of a man with me.

Wm. can talk of the latest poem in one breath and then expand on the glorious promise of Watt's steam engine in the next. He predicts that the Cornish miners will pay handsomely for efficient engines, as they are burdened with the high cost of coal from South Wales. He reports there is word about that Roebuck will pay off Watt's debts for a two-third's share in his engine. Watt will soon be on his way to London to apply for a Patent and Wm. has advised him to use general terms, which will give him greater protection from competition and greater scope for his own experiments. I know John Roebuck. He is a scholar of Classics who has surprised even himself by becoming a chymist of note, a member of the Royal Society and now an iron maker. Wm. has heard Watt's news and we are both delighted but confess to reservations regarding Roebuck, thou' why I cannot say.

Wm. sipped his tea and suggested we should adjust our estimation for Roebuck has made his forge on the banks of the Carron the finest forge in Scotland. I reminded him that it was the only forge in Scotland and made Wm. laugh.

I know nothing of the making of iron, he offered, but Boulton will indulge any man who asks him regarding the advantages of using coke from pit-coal instead of charcoal. It is a wonder he does not set up his own forge!

Of one thing, I am sure, I replied. He waits in the wings with his eye on Watt's engine. He plans to show Watt around Soho and the vast Manufactory he has created to tempt him into his own orbit if Watt's new Sun were to fade.

*

A.D. 1769 .... No. 913

Steam Engines, &c.

JAMES WATT

TO ALL TO WHOM THESE PRESENTS SHALL COME, I, James WATT, of Glasgow, in Scotland, Merchant, send greeting.

WHERE AS His most Excellent Majesty King George the Third, by His Letters Patent under the Great Seal of Great Britain, bearing date the Fifth 5 day of January, in the ninth year of His said Majesty's reign, did give and grant unto me, the said James Watt, His special licence, full power, sole priviledge and authority, that I, the said James Watt, my exors, admors, and assigns, should and lawfully might, during the term of years therein expressed, use, exercise, and vend, throughout that part of His Majesty's Kingdom of Great Britain called England, the Dominion of Wales, and Town of Berwick upon Tweed, and also in His Majesty's Colonies and Plantations abroad, my "NEW INVENTED METHOD OF LESSENING THE CONSUMPTION OF STEAM AND FUEL IN FIRE ENGINES;" in which said recited Letters Patent is contained a proviso obliging me, the said James Watt, by writing under my band and seal, to cause a particular description of the nature of the said Invention to be inrolled in His Majesties High Court of Chancery within four calendar months after the date of the said recited Letters Patent, as in and by the said Letters Patent, and the Statute in that behalf made, relation being thereunto respectively bad, may more at large appear.

Original Document: 1769 No. 913 James Watt Letters Patent

32

'his Patent in his Hand'

John Roebuck
Kinneil House
Bowness, 1769

I have had a timely visit, at last!

Black assured me Watt would ride across to Kinneil House and so he did, braving January's air, his cheeks flushed with the cold but his grasp firm in his handshake. It was a timely visit from this engine-maker who produced nothing but bills wanting payment. I suggested we go inside out of the cold but he said he was attuned to it these last few miles and would happily walk round my park.

I did not think him such a vigorous fellow but he walked steadily enough and seemed eager to discuss whatever I had in mind. You must relieve me of this Newcomen engine I began, it drives me to daily despair!

You must not blame the engine, he said, it can do no better.

But can your engine do better Watt, for that is the question, I replied, leaving aside my civility. Can you give me an engine which will produce three times as much power at a third of the cost of your Newcomen engine, I asked, adding, by way of a codicil, as you stated when we first met.

He stopped and regarded me. He spoke with a frankness that steadied my temper.

I can think of nothing but my engine. It is with me night and day but it is the scale, Roebuck, which holds me back. You have seen my little model work, but it will take more funds to reach the scale required.

I did not answer directly and Watt continued.

Besides, he said, we have water and coal aplenty and that is all steam needs to pump your colliery for a thousand years.

A thousand years!

Give or take a few hundred, he replied.

At that I did laugh. Watt is a man who can surprise one, one minute he equivocates, stumbling into uncertainty, the next he speaks in grand terms.

Of course, he added, looking at me with a forthright intensity, I should need the time and resources to solve the technical problems because few men can approach them with any certainty of the outcome.

But are you certain of the outcome, I enquired.

I am certain of the principles, he replied.

If so, such an enterprise must be protected by Patent, and I hope you have achieved that necessity.

I spoke with feeling, hard won by my own losses over my Alkaline works, where I had lost much for want of a Patent the Courts would uphold.

Watt nodded in agreement and further surprised me. He took from his coat a long paper with the telltale seal. With great satisfaction he handed the document to me.

I have Letters Patent, granted this week, he said, and stared at them as if he was even more surprised than I was.

He happily explained that on the advice of his dear friend and colleague Doctor Small, the Application was made on the grounds of General Principles rather than one or two particular engine designs. I held his Letters Patent in my hand but could no longer have my ears nipped by the afternoon air spoiling for the evening frost. I could not get us inside fast enough, to a warm fire and some Nourishment, for despite his bravado Watt was pinched and pale about the face.

We settled well enough, the Letters Patent on the table before us. Watt smiled across at me and further announced that he had acquired by his Patent full power, sole privilege and authority over all steam engines using his principles. All that remained, he assured me, was to inroll these Letters at the Chancery Court within in four calendar months.

I was hoping to entreat you to be one of my witnesses to my Lodgement, he asked, most courteously, his eyes shining once more.

It would have been beyond any man to resist and I could but smile back and assure him of my total support in his endeavours. We were sat in fine good humour with each other for many hours, like two contented cats sipping cream in the form of my best Madeira. As Black had predicted, Opportunity itself had ridden miles to shake my hand and I would not send it away.

*

33

'I am not a steam man'

John Smeaton has a growing reputation as an innovative civil engineer and physicist. He is a Fellow of the Royal Society and was granted the prestigious Copley Medal in 1759 for his development of a lift equation observed in waterwheels and windmills.

John Smeaton
London, 1769

As this is for myne eyes only it will take little application to write. Today was a day to remark upon: I am confounded as to whether it be for good or ill. In truth, I am greatly bemused!

I have received a plea for the commission of my time. A delegation arrived from the 'distinguished' New River Company in Islington along the Great North Road. They came to beg that I replace their horse mill with a steam engine. They arrived by appointment to my offices on The Strand and were much inclined to bowing. It took some time for their presentation to proceed such was their concern to honour my every word with a bow. They proudly showed me their records where the Company has been pumping water uphill for near on a hundred years into Claremont Square from New River Head, first by wind and then by horse. They are of a mind that the Company must be *up with the times* and have set their minds on steam power.

They unfurled a list of my 'known achievements' and reminded me, the supply of fresh water to the city is a civil matter in which I was the expert. I told them I am not a steam man, but they replied that I am a man of Water Power fame who holds a Copley Medal and they were confident I could make glorious improvements to their business and the lives of the good people of the Parish. They stood before me with their eyes wide and their pockets full of money.

Of course, there is not an engine-maker who does not know of the Instrument Maker James Watt and his Patent which he hopes to make rule these Isles. I am of the opinion Watt will never find a way to make his contraption work. Being an Instrument Maker myself I have not yet made a public comment never having given my time to Steam Power but I do not see about me the tools or the workmen to make true his design. In particular, our Ironmongers lack a proper boring tool and I hate to think what rudimentary rubbish awaits me when I am to make a steam engine for these good men of the New River Company. Even with these misgivings weighing on me, I accepted their commission for I could not bear another moment of their bowing and besides, I cannot be seen to be a Civil Engineer wary of a Civil Commission.

There was nothing for me but to provide them with a beam engine on Newcomen's old design, that is, the Common Engine. It is said to work tolerably well in many mines across the country but I had a dread that the configuration I used following Newcomen's principles would not suffice, and this proved to be true. The engine blustered about, but the operation made me aghast at its waste and output. It was barely fit for purpose and I cannot let it stand.

I will now, for my sins, become a steam man and put my mind to Newcomen's giants in other places and see what figures I can amass to guide me. I have vowed to make of this giant a true servant for its owners. I should laugh at my own nature, which has me vowing This and vowing That, like a horse driven hard by his masters Cost and Efficiency. They rarely let me rest and have me pulling all sorts of loads. Now I will pull one for the New River Company and pray I can do so at a distance from their bowing.

*

34

'Of all things in life there is nothing
more foolish than inventing'

James Watt
Kinneil House, 1769

My days are much the same; I rise in Hope and retire in Disappointment. Peggy, dear girl, writes that I must not despair but despair has become like a dog snapping at my heels. She comforts me with news of the family and her confidence in me. I find myself falling into a meditation where she wipes my brow and I smell violets on her hands and sleep is upon me. Such fancies do not last a moment before the world is upon me like a battle drum in my ears.

I have this day received a note from Wm. Small who has become a most devious fellow and goads me well. Last year it was stories of the indefatigable Irishman Richard Edgeworth (1) and his enthusiasm for Doctor Darwin's *Fire Carriage* and Doctor Darwin's enthusiasm for Edgeworth. As a warning I might be gazumped in the making of a steam carriage by Darwin's new Genius, he writes I may expect such a contraption to come tumbling down the road at any time! He does not comprehend that I am now so spent I can barely raise interest in reading gossip! Besides, if Edgeworth could bring forth Darwin's fantastical *Fire Carriage*, he is welcome to any glories it might bring beside the terror of the populace!

Since Roebuck and I are signed business partners there is no peace in the world. Small and Darwin have been in my ear like ghoulish shareholders. Small asks if I have invented any new gimmicks lately as if he expects me to produce one a day like Wedgewood's bowls!

They both write of the future of Steam with great optimism and seem to endow me with skills and endurance I may not possess. As for steam carriages, I do know this – we have not the knowledge or the tool makers to create such a thing and my efforts will go to what is known and already prov'd: a steam engine pump.

Roebuck has settled my debts for a two-thirds share in the engine's profits and provides the funds necessary for its completion. He is driven to distraction by his flooding problems and the inadequacy of the Newcomen Engine although I pointed out, in defence of the labouring beast, that it was not in the Engine's design to do better. He has me back at lodgings and a work shed at Kinneil House which he says is convenient to his mines at Bowness. He has my workshop, as he says, hidden from the eyes of those who may seek advantage, deep in his park. I am become a secret to the neighbours and to myself. Thankfully Peggy does not see the state of my clothes at the end of the day when the light is too dim even for my fevered brain.

I have foregone my experiments with inverted cylinders and given my time to the beam engine with a condenser. Its design will no longer lie idle in pretty inks on my Patent. Roebuck agreed my work must be to full scale and I set about the engine with great hope in my heart. Through April, May and June I have given myself entirely to this work. I was not to know what tryals lay ahead and, had I even glimpsed what lay before me, I might have given up the whole enterprise. I am fast believing that of all things in life there is nothing more foolish than inventing.

Roebuck says I am in a Brown Study and he cajoles me that a Day, a Moment, ought not to be lost, as if I am withholding Success from some malignancy of mind suddenly upon me. There are two frustrations on my progress, the first being that Roebuck's men can smite iron well enough but they cannot bore a cylinder with the precision I require. The second is that I can find no way to make the piston airtight. My design does not allow for a layer of water above the piston as Newcomen did so

I am having a devil of a time making my cylinder steam-tight, that is, air-tight. I thought an oil seal might do but under pressure it became thick and clogged the pump. I understand oil to be a wondrous substance with great versatility but I am now of the belief that oil will only act as a sealant when the differences in the measurements is minute and hardly visible to the human eye. Every day a new solution springs into my mind, felt, leather, hemp, all to no real purpose. I have persevered in the face of these difficulties, but it is not inaction which makes my head ache and my hands shake so I fear I may be in the grip of a palsy, it is the realisation that Endeavour which cannot recognize Defeat is pointless.

Small, thinking of my family, now suggests various surveying positions but I am fast ahead of him and have procured a contract for the laying of a canal through Strathmore. True winter will be upon us soon and I dread the lost days at Bowness Roebuck will face or the long hours ahead of me in all weather, usually the coldest when we find the most unforgiving soil set to break any pick we may kiss it with. Peggy writes that I am Blessed with many talents and I should turn to such Inventions that please me for, she reckons, without any measuring stick but Love, that I should have at least a hundred left in my journeyman's bag.

It is true I am already thinking of some useful instruments to adjust a survey level and to measuring distances. In particular I thought a telescope with adjustable cross-hairs in the eye-piece could be useful for measuring distances, especially between hills and across water. This notion of what I will call a *Micrometre*, if satisfactory, would certainly aid any surveyor caught in a squall laying chains on the ground with bleeding knuckles too cold to feel. That may well be, and a good thing, but whatever activity takes my mind, it will be but a momentary respite for I have yet to bring life to my engine left idle – idle, and rubbing at my soul.

*

35

'What can be done with one small engine
can be done with many.'

John Smeaton
Austhorpe
West Yorkshire, 1770

Austhorpe House is a draughty monster that grinds its footings on a windy night as if to tell me it might one day take flight like a Dodo and ruin my reputation. I have held off its Refurbishment hoping to replenish my funds from steam commissions which grow apace. Fortunately, it is happily removed from London and quiet. How a man can think in the clamour of a city I do not know! I am thankful to have my family about me, my precious daughters, and my worthy assistants who are learning by and by to follow me. There is space and time enough in my tower workshop to make any experiments I wish. I have a forge and a good lathe and will make progress if I have the time to think on it. I have commanded of myself the habit of working undisturbed by any who would break my concentration. Even my blacksmith has learn'd to wait!

And now I am upon steam. My next engine will be more advanced in both power and economy than any previous Common engines which may be timely as I have heard that Watt is spending his time surveying and the world has no news of his engine.

There are masses of notes in front of me which my daughter Ann, who is as inquisitive as a squirrel in spring, calls my spider notes, taken from every working engine in Newcastle and Cornwall that I could get my men to measure. They speak to me in the clearest language an engineer should know – Measurements and Measurements. They are my stepping stones across the slippery pond of steam. They will provide any man who seeks to know with *A Table*. This is a Table which I will publish as one which can be reliably used to gauge the properties for engine. I will set it at first for cylinders of 10 to 72 inch diameters. Moreover, to make a standard to follow I have set what I call a 'Duty' to measure the performance of each engine. The 'Duty' measures the amount of water in millions of pounds that the engine can raise by one foot per one bushel of coal used.

There are engine men who will say such a Duty is far too onerous a calculation and I can hear their wailing across the shires already, but when it is set and repeated especially as it focuses on the essential task of a steam engine - the raising of water - it will become, it is my hope, a standard practice for all to follow.

Now I have a Standard Duty for myself to follow, I have been surprised to find that the size of the cylinder does not reckon the efficiency of the engine. Between differing engines, I found the performance varied greatly so that a 60 inch diameter cylinder performed better on one engine than another having a cylinder diameter of 75 inches. What I suspected about our tools for manufacturing was revealed by the discrepancy in the degree of accuracy of the components. I have marked the following:

1. If the cylinder allows the piston to be slack in the bore performance will be less.
2. If a valve does not admit the right amount of steam the engine will suffer.
3. If the stroke is too short the engine will suffer.

Following these observations, I have found simple adjustments can improve performance. For instance, it is important to measure whether the grate of the boiler is close enough to provide the right amount of heat. It is necessary to check that the amount of water on top of the piston is regulated to stop excessive cooling.

My engine at Austhorpe has a 10 inch diameter cylinder and a small performance. It is a sturdy little thing and I have given it over 100 adjustments. These adjustments were my guide and by constant testing, varying one component at a time and holding all others constant, I have improved its performance.

I have made a new piston with a wooden underside instead of the leather previously used. The state of a leather sealed piston, cut and leaking, was a horror no engineer could tolerate. To my satisfaction I have increased the flexibility of the working strokes by regulation of the volume of steam. As the engine load decreases because there is less water to pump, a cataract is used to regulate the number of strokes per minute. This cataract adjusts the amount of water held in the cup to achieve a variable weight effect. To add to my adjustments, I have ordered a hemispherical cylinder bottom and set the beam's trunnions or pivots at its midpoint instead of below as in the Common Engine. I have in mind to strengthen the beam by the use of fir wood in a laminated process. This is slower because it requires the wood to be sawn and bent, but it will add to the strength of the beam. The task I have set myself is clear. The next engine I build must lift a maximum weight of water in a timely manner and with the least amount of coal. It will then be true that what can be done with one small engine, can be done with many.

*

36

'value the engine as less than a pouch of farthings.'

John Roebuck
Bowness, 1772

I have thought these many weeks of my mother and in one particularity, and that particularity is much upon me.

She was a country girl who married well or so she used to boast. She would laugh when she said it as if it was the greatest joke in the world. She used to talk about her brother who went to sea and on his visits home would take her on walks across the wild ridges of Yorkshire and tell tales of his adventures. One such story was of a Fish who was asked, what News from the Sea, and the Fish replied, I have a lot to say, but my mouth is full of water.

She told me that story many times and perhaps she thought me able to explain it further for I too had traveled if only north to the mists of Edinburgh. It is a doomed thought that this memory returns, for News from that city has left my mouth full of water as if the wind has carried the Northern Sea up the Forth and seeks to drown my very life away.

The life which was in the Ayr Bank. (1) The bank which fed so much trade and enterprise with its credit, and upon which my fortune rests, has fallen! Despite what unnatural sympathy is wasted upon the partners, I am one who can hardly bear to write their names, but Fordyce (2) is the greater villain. There is Pity that such men have fallen, but there should be more Pity for those whose money fell with them. I cannot believe that I am now bankrupt.

We will lose our holdings in many enterprises, for my wife and I put our own fortune to work the mines at Bowness and maintain our partnership at Carron. Scotland cannot let Carron fail, for it is the livelihood of thousands! My friends have rallied in concern, but concern alone will not save the day.

The whole country is aghast from London to Edinburgh as the state of Fordyce's misadventures are revealed. It is his foolishness which has caused our Banks to fold one on the other as a house of cards. Stories circulate that the partners traded beyond their means and gave credit too easily, such obviously imprudent behaviour has made my guts twist and heave and I would be disinclined to give these partners who lost their entire estates much mercy. All about is melancholy and tears and grown men are said to weep!

There is Loss of Credit in all directions, and suspicion of undertaking any business is rampant. It will be the hardest business to tell Watt I have nothing to give him towards his Engine, yet I hear he is asking others to offer to buy out my share for my sake!

There are mornings when I am so downcast I cannot rise from bed or lift my head even be it to curse Scottish Bankers. My wife has such a rage about her she has become a Warrior Queen taken to keeping our Creditors at bay. She is Fierce against Fordyce, whom she is calling a Charlatan of the First Order and, I fear, were she ever to be in the same room as him she may well end up at Tyburn where Fordyce himself should dangle! She insists I will rise again in other endeavours. When I see her determination, I do not doubt I will take on some new Enterprise, but my association with Watt's engine and the promise it holds tears at my heart and in my melancholy I pick at it like a sore scab.

*

It is four years since Watt and I came to our agreement on that cold January day, full of the enthusiasm necessary for greatness but at odds with the time about us. To support his family Watt has taken offers for his services as a Surveyor, which led him to endure the hardships of the wild country where nothing short of a fortune would have enticed me to go. We are both men with families and I know Watt will understand that I have taken Boulton's offer, for he is one of my Creditors and I have handed him my share in the Kinneil Engine, in lieu of an amount to £1,200.

My other creditors looked upon the engine as next to worthless, counting its value as less than a pouch of farthings. Its fate was of no interest to them. But not so Boulton. Some months ago dear Watt had pushed for Boulton to be offered a share in three Counties but he had a high disregard for such pickings and declared he would rather make engines 'for all the world.'

I should bow to Boulton for he and I have long known that Watt's engine offered a money-getting, ingenious project and he has bided his time until Lady Fortune swung his way. I will say this for him, he is far more than a Button Maker. He is a leading light in the Lunar Circle, men whose confidence fuels a vision far beyond the limits of ordinary men.

Even though my wife will not let me say it, I am now an invisible man, left adrift from great enterprises. I knew as I signed over the Kinneil engine, deep in my bowels, that I signed away a great fortune, like a seam of gold lying close to the surface near men with ready picks. Boulton has already had it dismantled and sent to Soho and I know he will find a way to entice the engine-maker to the engine. Watt will, sooner or later, fly into his orbit for Boulton cannot let such a shining star escape if he is to make engines for all the world.

*

37

'A Fate worse than Death has come to me'

James Watt
Dunbarton, 1773

I am in despair so deep I would call the world profane and all its Beauty dulled forever in my eyes, for I am in dispute against my own breathing body that it persists to have me still alive.

It has been cold for days and chilled me to the bone, and my ears sting with hot needles driven by the wind that stunts and sours every tree and hedge that it can find to worry. Dear Hamilton has sat me in an Ale-House against a good fire and forced Brandy into me but there is no heat that can reach the ice around my heart.

It was dear Peggy herself who had me go again to Survey, even though she was with child, much to our joy, because it was better for me to be gainfully employed than home fretting about the Engine she declared and kissed me so sweetly goodbye! We parted most happily for we both called the coming child a welcome gift to our family and prayed only that it should be healthy. There are stones still that knock in my heart from the death of our firstborn John and shortly after the loss of our angel Agnes, named after my mother, also lost before her time.

Now Hamilton brings the news that Peggy is lost to me and the child too, which likely was the cause of her death. And all done days ago with me not by her side to offer her comfort as she has done so many times to me. Hamilton is the dearest friend, for he has come from Glasgow to meet me on the road from Fort William and offer a condoling presence for I am bent as a cripple might be with grief upon my shoulders and cannot think of anything sensible beyond a minute.

It comes wailing at me that I should never have gone, leaving her frail with her own grief and with child as well as Mistress of the business and mother of two. She was such a good wife and I a poor husband to leave her so.

I waited quietly by the fire, for I wanted my bones to crumble this instant into a white powder so they might mix with the ash about this hearth and my existence be swept away by the scullery girl in the morning. But Hamilton would have none of such sitting. He fed me more brandy and said I would be better for a night's sleep. He has offered that I can stay with him for the present as I am unable to face the prospect of home in any way that would be fit. That I will not hear her voice in welcome is quite unbearable to me.

He has suggested that I write as soon as possible to my colleagues and friends, who, he assures me, will offer me great support. I know that is true, especially of my learned Lunartick friends, but I know the advice that they will offer will be the same advice that my father, dear man, gave me at the death of my mother, then for my dear brother lost at sea and for my dear sister lost to a coughing illness. He said the same: unless we prostrate ourselves to chill to death on the stones of an altar or weep a thousand tears or let our body roll and rot in the Clyde, we have no other choice but to get to our work and let the day, soft or hard, unfold as God wills

That is what I would say to any man, were he to ask, but whether it be comfort or not depends entirely on the man. I would have it comfort me if it were true, but as I write my schemes and gadgets seem trivial and mere child's toys where it would not matter a pinch if they be left idle to the End of Days.

I will lodge with dear Clive Hamilton's family and find the time to write post haste to Wm. Small, Black and Darwin. I will put it to them with Great Urgency that something must be done. The great minds in the Medical Faculty at Edinburgh must to do better with their studies for we cannot have a civilised society that tolerates its women dying so readily in childbirth and losing infants not even blessed with putting a foot upon the Earth.

It is intolerable that the Universities are awash with medical students and Professors who gather their Observations to make their names in publishing, yet my wife dies with a doctor entirely unable to save her!

I watched my father lose half himself when my mother died, yet he picked up his tools and went to work and I took it as heartless. He did the same when he lost his two grown children and has lived his three score and ten, yet keeps himself busy even now because of me, his last living child. I have two alive to call me father and when I pray tonight I will tell Peggy not to fear for them. My body aches and my hands shake but Hamilton will be with me.

To comfort my soul I wrote to dear Wm. on my miserable journey home where the roads were rutted streams in the incessant rain. It took three days and I was as wet as water could make me. I could scarcely preserve my journal-book and did not know the difference between my tears and God's rain. My heart was a heavy stone and my horse and I could barely carry it.

*

It has been months in which I have given myself to grief, as Wm. suggested, with my devotion to work. I have had my head in calculations and my body out in all weathers surveying on the Firth of Forth battered by a south-west wind with a tongue of ice. I am not the first man to lose his wife in childbirth but I am still reduced to darkness and care not for things that should be dear to me. I am so poor I am unable to further my desires nor have the desire to do so. I have written to Small that my country repels me and I am heartsick scraping a living and fearing for my future.

I do not know what to make of this but Boulton has called for me to come to Birmingham, where he says the Kinneil Engine could well do with my attention should I be so inclined.

*

38

'that he would promise me the Sun, the Moon and the Stars'

James Watt
Birmingham, 1774

I arrived in Birmingham and noted the green shoots already on trees and the May Blossom in bloom with no feeling left in me, for I had turned my back on my own country with all its sorrows and had nothing but threadbare hope for what lay ahead. My father quoted from his reading. He said that no man should do nothing because he thought he could very little and told me of this with his hand in mine. So it was I left Glasgow. I promised Maggie and Jim that I would send for them as soon as I was able, and knew my father's care was not far from his grandchildren. They watched me wave goodbye, Maggie close to little Jimmy as I had instructed, their faces as forlorn as my heart.

Boulton met my coach and escorted me to his old home at New Hall Walk which he announced was available for my new lodgings if I was of a mind. It is down a slight slope looking out onto green fields with a carriage way defended by young elms. I was half in expectation that he would promise me the Sun, the Moon and the Stars as well, and so he did!

I was in no way of being amused but he did calm my sorrow. As soon as I was rested, he had me dine with him at his family home in Soho at a table so elegant and congenial I felt most comforted. Yet when he spoke of success I became sick with apprehension. As if he were reciting news which had already happened, he told me of the Engines we would build and the countless mine owners who would pay us extraordinary fees for the efficiency of our designs. He said his partnership in the manufactory with Fothergill would not deter a new partnership but complement it. He swore he knew mine-owners as far afield as Cornwall and Yorkshire who were waiting on what Boulton & Watt could provide for them. He spoke of the Kinneil engine with great fondness and mentioned he had already had ordered it unpacked, cleaned and oiled ready for my attention. It will be put to work, he said, for that is what engines must do.

I was silent for some time and Boulton, most kindly, seemed not to notice. I was determined not to be wooed by his generalities however shiny their coats. Before me was his silver-plated dinner ware which I had been admiring but which appeared far from my own achievements. I did answer, finally, and my melancholy gave me a wild strength I did not think I possessed. I spoke of the mountains I must climb and the costs I would incur. I spoke of the horrors I had in procuring cylinders well cast and most not even half suitable for their task.

He slapped his thigh, most pleased with himself, as if he could answer such difficulties. He told me of his great friendship with the iron genius John Wilkinson, whose foundries at Bradley and New Willey he had personally visited. He described their sand pits lined with loam and how he had seen the fiery streams of molten iron seething sparks and fury. He spoke in awe of these iron men who sing lusty songs as they stand amongst the fire devils whirling at their feet and clear the mould vents of gasses. Boulton declared that Wilkinson would, more than any other man he knew, make iron fit for great constructions beyond even our desires.

I do not know how much you know of Coalbrookdale, he said, not waiting for my reply, his eyes gleaming with the import of his declarations. I would not be so certain of our success without them, he continued. Scholars of history will talk of these men of iron, for it is their innovations which will change the world. It is not so long past that the Quaker Abraham Darby Jr devised a way to improve the production of wrought iron by coke-smelted pig-iron. And not so long past that the Cranage brothers devised a reverberatory furnace which heats on all sides and does not use scarce charcoal, but plentiful coal.

Their furnace allows the iron to be affected by the heat of the coal to undergo conversion, but happily does not suffer by mixing with it. Boulton paused to take breath.

I was about to respond when he replenished my glass and gave me the most warm and unexpected smile. At every turn he was most convivial and in the shortest time imaginable we were in agreement that he would bear the costs well beyond my resources but necessary for our success.

The more I detailed the expenses the more he agreed, seemingly pleased with my forthright demands. It was obvious he had given the whole venture much thought, and declared that time was currently our enemy and that we would need a surfeit of it to reap the rewards due to us. Under such conditions we will need to extend your Patent he announced as if he expected delivery of such in the morrow's post.

It was well beyond my expectations and I wanted suddenly to laugh! My spirit stirred as if awakening from a dreadful sleep and my blood was warmed through. I saw that Boulton was more than an Adventurer looking to make good on his investments. I sat with a man of Insight and Intellect, who did not allow his dreams to burden him but let them run ahead so he may follow and take others with him.

You must experiment as much as you desire and I will bear the costs of such and the workmen you deem necessary, he added, waving a hand through the air as if to conjure away my failures at Kinneil House. Time will be our only limitation, Boulton said, and what was once thought impossible will be possible. He leant across the table and raised his glass. His cheeks glowed with good humour and he said most heartily, the world is waiting for us James Watt - we must not disappoint!

***

REVOLUTIONS

PART 1

Steam Team 3

John Wilkinson 1728 - 1808

Matthew Boulton 1728 - 1809

James Watt 1736 - 1819

Matthew Wasborough 1753 - 1781

William Murdoch 1754 - 1839

39

'we drank from the Cornucopia of Hope'

Erasmus Darwin
Lichfield, 1775

It has been a poor lamentable winter, the days long and the nights longer. It has been a month since I watched William Small draw his last breath and saw Death's hand smooth his face to boyhood. It was Boulton who closed his eyes and offered a prayer for I sat so abject I was unable to move. If he had been my own blood I could not have been more affected, for he only lightly wore the disguise of a Colleague and Philosopher, being in truth a brother to us all. I cannot recall a time when he has not been there to offer the fruit of his large and liberal mind, his generous attention to our tryals and tribulations, his time and funds when they made a difference to our well-being.

Boulton has written to our Circle and called a meeting at Soho House next week on our Worm Moon. It brings me a bitter thought that his haste is fuelled by the possibility that his Engine-maker may take flight to Russia for Watt is forever counting his pennies and the Imperial Government has offered him a lucrative post. We have only just lost Small to God and now it seems we may lose Watt to the Russian Bear and all the wild country in between.

I have heard, in my desolation, dear Polly's voice of late, God Rest Her Soul. She would say it is high time our Lunar Society met and being a company of Small's dear friends we might well comfort each other. She was always right in such matters and when Watt agreed to share my chaise on the last part of my journey my despondency lifted.

*

After a full day's travel, I have returned late as usual from Soho House with a clear night sky for my company after Watt was deposited at New Hall. I would have scoffed bitterly if an Angel had predicted I would ride home with Watt in a fine good humour. But so I did, chuckling from time to time and jingling the coins Boulton had given me declaring Minting to be his new interest and his intention, in the fullness of time, of saving the Coin of the Realm from counterfeit rascals.

Our journey towards Soho had been a miserable affair, with Watt his usual damp squib of a self, poor fellow, complaining of all his aches and pains until I threatened to stop the chaise and strip him naked for further inspection of his ailments. He had the grace to laugh and told me he was glad I was Boulton's doctor and not his own. He then confided in me his fears over the fate of his Private Bill for his '69 Patent extension, given the prevailing winds against it. I agreed that a Rebellious War in the American Colonies was a challenging wind to overcome, but assured him that he had many friends and we would do our very best to capture the votes of Members as they waited in the Lobbies. War or not, I concluded, there is not a worthy man in England who does not want Commerce to succeed and you are the man who will drive its Success.

He laughed again and said, you are a fine tonic, Erasmus, and have quite warmed my soul.

We arrived on the stroke of two of the afternoon by the house clock and were shown into Boulton's dining room where the table was set for light refreshments with fresh quills and paper aplenty. Boulton, being the host, called us to order as some lit pipes and others sought cushions and leg rests for their comfort. Including Boulton, Watt and myself; present were Josiah Wedgwood, Joseph Priestley, James Keir, William Withering, Richard Edgeworth his friend Thomas Day.

Boulton began very seriously. We have tonight a Paper that will be given by James Keir who as members know has travelled from Top Hill where he has been working on translating Macquer's *Dictionnaire de Chymie*, which is well beyond my French.

The members tapped the table in robust agreement as it was well known that Boulton's French was deplorable.

Tonight we are privileged to hear his recent paper written especially for this occasion. It will be *On the Latest Crystallisations observed in Waistcoat Fashion*, Boulton continued.

Watt, who was my neighbour at the table, nudged me as if to get my attention but I was already alert and about to query our Illustrious Chairman.

Watt spoke before I did.

Mr Chairman, he said, I am at a loss. Could you please repeat your last sentence?

My last sentence? Boulton responded as if surprised. I do declare, Watt, those engines of yours have been thundering in your ears and rendered you deaf. Did you not hear me say the paper is to be *On the Latest Crystallisations observed in Waistcoat Fashion?*'

Several members tapped the table in support of the Chair as that is exactly what they had heard.

Before I knew what I was about I was on my feet.

I am at a loss also, I declared, for the observations of Waistcoat fashion, while of interest to some no doubt, I'd wager none of those sort are among us this afternoon.

Good Heavens man, said Boulton, not interested in waistcoat fashion! I have it on very reliable report from your Housekeeper that you have been known fuss over your waistcoat as you dress.

Since this was very true and known to many present, I stood completely nonplussed while the members tapped in some merriment on the table. I sat back and found I was without words to counter Boulton, who seemed to have lost his senses.

Wedgwood then rose to his feet saying, Boulton my good fellow, I have no idea what is going on, but if I could offer myself and my waistcoat as empirical evidence to support Keir's paper, I would be a happy man.

I heard Watt splutter into his ale as if he was attempting to drink and speak at exactly the same moment.

Thank you, Wedgwood. Could you perhaps display your waistcoat?

Before Wedgwood could respond, Priestley sprang to his feet and, flinging his coat open, announced his entrance into the fray.

Mr. Chairman, I beg of you, my waistcoat is far more of the fashion moment than Wedgwood's being a Radical design, to which the assembly thundered on the table.

Wedgwood parried immediately. How can that be so, he argued, when my waistcoat is covered in daringly stitched green frogs to celebrate the Russian Empress Catherine buying over nine hundred pieces of my Green Frog Service?

Gentlemen, gentlemen, said Boulton, a smile spreading across his face. Your disagreement is a Small matter.

I rose to my feet again, most unsettled at the madness which seemed to be growing around me.

A Small matter you say, I stormed at Boulton, what nonsense is this?

Why Erasmus, you look as though you are about to call me out on this Small matter, Boulton replied.

I felt Watt tugging at my coat. Erasmus, take a look around you, I beg of you. Watt had a strange but not unhappy look.

I then saw that every man had moved his coat aside and was displaying his embroidered waistcoat. I turned to Boulton and cursed him roundly, for there were no ladies present. I did not know whether to laugh or cry but since my fellow Lunaticks looked so pleased at their jape, I decided on the former.

My Lunarticks, including Watt, were wearing a most exquisite waistcoat in honour of dear William, who had loved his waistcoats. It was then, as I stood gathering my senses, I saw that I was wearing my latest waistcoat laid out for me by my attentive Housekeeper, who had fussed greatly over my appearance. I had tears in my eyes, as did a few of us, as I spoke most respectfully to Keir, who had been waiting patiently throughout.

You are the most wonderful Lunartick, I said to him, and I await your paper with great interest. With that I sat and drank the heartiest draught of ale ever to pass my lips.

Keir did not disappoint. His paper *On the Latest Crystallisations observed in Waist Coat Fashion* was a fine homage to dear Small capturing his keen intelligence, his wit, his foibles, his desire to find affinity not only in principles or compounds but in his love of his fellow man. Keir described in detail his warm kindness to us all, in story after story given to him by those present.

Boulton suggested every man should be well irrigated and we laughed and sang and shed tears and became full as goats, and when dinner arrived we were as ravenous as wolves. I was never in my life so young. The empty urn of Grief was discarded and instead we drank from the Cornucopia of Hope and all things were possible once more.

*

40

'I am content enough'

John Smeaton
Northumberland, 1775

I hear Watt is in Birmingham but do not know what he is about. There is common talk that he has taken up with Matthew Boulton of Soho and I cannot imagine how Watt will organise his mind in such noise and bluster which surrounds Boulton. For myself I am content at last with the engines at Long Benton and Chacewater mines and am free to consider other commissions of my time which will relieve me of the need to stand by hot, pulsing steam and fret that I may blow myself and all about to Kingdom Come.

There are water mills and flint mills and harbour channels aplenty. Stone and brick and mortar are sensible materials that leave men calm about their business, but steam breathes and snorts and pretends all manner of moods until the poor engine-maker runs hither and thither at its beck and call. Besides, I have stood quite long enough in the winds of Northumberland, shouting through incessant rain to workmen who cannot remember their instructions from one day to the next.

It has taken two years to bring both engines to a starting point where I have them working at greater efficiency than any steam engine thus made. I would not boast this claim in publick but I have notes to prove it. At Chacewater I have replaced two engines with one, and given it a 72 inch cylinder to increase give power. I have measured an improvement of 25 percentage as what was done in twenty-four hours will now be done in nineteen hours and twenty-two minutes. The coal consumed was eleven chaldrons a day and I have calculated the new engine will use seven and this with the same amount of water lifted. The Long Benton engine is working well with a 52inch diameter cylinder that stands at 6' 6" and has kept airtight; the working stroke will be 5' 8" and will take 14 1/2 strokes a minute.

This increase in power will become known to Watt if he cares to lift his head and look about him. It is also said he has been Surveying, leaving his Patent to hold the field for him if he wishes to return. I am almost of a mind to write to him but I will desist for I have heard he is greatly diminished since he lost his wife and child. When he is of a mind to show his wares, the world will stop and consider, I do not doubt, but meanwhile no man can progress steam power against his Patent now extended to 1800, if indeed, any more progress is possible.

There is a particular satisfaction that I may claim. It is for the new cylinder I used, bored at Carron Works by the new boring mill I sketched. It is said that 'A bad workman will never find a good tool' but never was there a more unjust proverb for the Ironworkers. Strong men have bent their backs yet weep at the demands of a design they cannot meet for they have not the tools or instruments for accurate execution. My boring mill will help to produce a truly circular bore, which will aid all manner of things, some yet to be imagined such is the pace of invention across these Isles. But it remains true, any man who can cast true engine parts will be a hero to us all.

*

41

'Boulton does not shrink an inch'

James Watt
Birmingham, 1776

It is March and the weather is unsettled but that was of no consequence at Bloomfield for the engine paid no heed to the weather, only to the Fireman who stoked its furnace and the coal boys who ran its coal.

Boulton pushed for this moment, and he stood in his best satin and glowed in the reflection of the praise of all. He wanted me be satisfied with every part before we began, for the sake of an Occasion to be Remembered by All as wondrous and profound. He wanted the Investors cheering and the Men of Science stilled to Awe. So it was, for the engine began in a great hiss and rumble and the first return sprayed a veritable vomit of brown, turgid water into the draining pit.

It was the first such moment for Boulton & Watt and had I not been given to anxiety which plagues me dreadfully, I would have cheered with the workmen. I waited hot and bedevilled until we measured the strokes. It was a relief to hear they were at 15 per minute, steady, Mister Watt, as the boys called to me. They were as happy as I was and made me feel that I had built a fine Engine and that it was behaving itself and determined to make us proud. As I listened to its rhythm I knew its birth was of good stock. The cylinder was by way of Wilkinson's new Boring Tool at Coalbrookdale and the valves, plugs and condenser had been made at our workshop and I began to hope that at last our Workmanship and our machine Tools could keep pace with our desires.

It has a 50 inch cylinder, the largest we have made, and took a minute short of an hour for the engine to pump some 60 feet of water from the pit, to the great satisfaction of the owners, the Messrs. Bentley, who immediately invited Boulton, Harrison and myself to dinner and a worthy celebration of the new 'Parliament Engine' being so named by the workmen as is the custom.

This being the first of our engines *Ari's Birmingham Gazette* had sent a man and an artist to capture the scene, no doubt at the fulsome Invitation of Boulton & Fothergill. Boulton himself was in his Element. He declared himself 'hot with Port' and insisted on taking my arm and introducing me to whoever crossed his path as 'his Engineer'. This was Boulton's triumph; he does did not shrink an inch from any endeavour while it is clear that I tarry too long and forget that action can solve a thousand problems.

It was with some foresight that he had ordered several carriages with footmen to transport us back to our Lodgings for it was a sad truth that men from Boulton & Fothergill's manufactory, including at least one of the Principals, were struck with a sudden malaise in their legs which rendered them unable to walk but very inclined to sing verses their wives would never have thought known to them.

It is eleven long years since my little assembly with a separate condenser showed me the way forward and today our engine worked more than tolerably well. As I returned to see how the lads were managing, I stood and regarded the whole device. Its parts moved in union, smooth and steady and if it had been a giant cat I thought it might purr at me it was so content. Roebuck, being a fine chymist, might have told me what it was I had about me, a smell of warm iron and brass and probably phosphorus and sulphur and being so familiar I was happy to be warmed by it.

I could not move away for a cloud passed and the yellow moon spilled its light into what my mother would say was every nook and cranny the eye could spy. The moon was my inspection torch and it guided my thoughts. I could clearly see that our chain linkage holding the piston rod to the beam is well enough for pumping water but it was not enough. I determined in that moment, that any engine should answer to pushing, as well as pulling. Presently the action of the beam affected the pump rod and the piston and limited the movement of the whole. Something must be done, I thought, to divert to a straight-line motion for them bothe, but I was full enough of good ale to laugh at myself making plans in the moonlight.

When I turned back to the Celebration another consideration entirely came to me. I thought of Ann McGrigor in Glasgow whom I had recently visited. Surely a hen who may want to flap her wings in wider fields than her father's house, and I wondered too, if she would have me.

You have ideas enough that they will wait for the morrow I could hear Wm. Small's calm observation as if he stood with me just now amidst the huff and puff of success. I would have him here to witness our success. I would take his arm and introduce him to whoever crossed my path as My Excellent Mentor.

*

Ari's Birmingham Gazette
11 March 1776

The Engine was made under Mr. Watt's and Mr. Boulton's Directions at Boulton and Fothergill's manufactory near this town; where they have nearly finished four of them, and have established a Fabrick for them upon so extensive a Plan as to render then applicable to almost all Purposes where mechanized Power is required, whether great or small, or where the Motion wanted is either rotary or reciprocating. A Number of Scientific Gentlemen whose Curiosity was excited to see first Movements of so singular and so powerful a Machine; and whose Expectations were gratified by the Excellence of its performance. The Workmanship of the Whole did not pass unnoticed, nor unadmired...

The liberal Spirit shown by the Proprietors of Bloomfield in ordering this the first large engine of the Kind that hath been made, and in rejecting a common one which they had begun to erect, entitle them to the thanks of the public; for by this Example the Doubts of the Inexperienced are dispelled, and the Importance and Usefulness of the Invention is finally decided.

Original Document: p *Ari's Birmingham Gazette* 11 March 1776. Boasting the success of local manufacturing.

42

'Let me not to the marriage of true minds admit impediments'

John Roebuck
Kinneil House
Bowness, 1777

My wife says she has become greatly perplexed. She says I live in the skin of a chameleon and reminds me she married a physician who became a chymst who became a manufacturer who became an iron master and is now a farmer. She worries she does not know what I will become next and has taken to the pages of *The Scots Magazine* (1) to find out!

It is true I have taken to farming at Kinneil House and complain that the tenants are slow to plant trees and find them stubborn against feeding their good turnips to sheep. It seems a lifetime since I battled at Bowness. I look back in fondness at our tryals where our miners were thirsty for change and readily turned to any enterprise asked of them. It may be easier for me to fund an expedition to move the Pyramids of Egypt than ask a Scottish farmer to change his ways!

And now we get word from Birmingham where they talk of nothing else but the success of Boulton & Watt. I predicted Boulton would be a great match for Watt and quipped to Ann, that it would be a Marriage of True Minds, but I am amazed to see how quickly they have progressed. It presses me tho' and the Devil pokes my disappointment which cannot last the hour for Watt himself is a good fellow and friend. He released me from debt and continues our association, reminding me that I was the first to invest directly in his Engine and without me he might have given the whole struggle away.

We hear he has married Ann McGrigor of Glasgow, which is a good thing for he was much reduced and we feared he would be swallowed by Melancholy and disappear from the world entirely. Instead, he has been taken into the folds of the Lunar Society and has friends and connections who stand in solid fraternity. It is now certainly true they are fast becoming the most powerful philosophers club in England. The Royal Society may have the prestige but these Birmingham men are determined to make their mark and do not worry themselves about the Niceties of their actions. The procurement of an Extension to Watt's Patent by way of a Private Bill to Parliament is nothing more than a brilliant piece of strategy. I can well see Darwin, Wedgwood and Keir in fine good form around the smoky Lobbies of the Commons gathering votes! This Bill includes Scotland and gives them protection until 1800!

I am still in Admiration of their Temperament and count them all as friends. I have sent them news that I have joined the Lords and Lairds of *The Improvers*[2] and hope to bring science into the mud and sweat of Agriculture, for any educated man could do no less having witnessed the poverty of the cottars and their manner of Subsistence whereby their children, like their animals, struggle to thrive.

That Agriculture with all her needs has come under the sway of ready minds is a great Blessing for these Isles. Dr Darwin has often suggested there is a happy, nay, natural connection between plants, animals and humankind and, in fear of becoming 'mystical', I feel a bond with the land approaching such a union. But the smell of earth on my hands will not keep me from shrinking in envy of Watt's success.

I knew envy stalked me when I counted the twelve years since I was made a Fellow of the Society. No matter how many times I count it is only twelve, but it seemed a lifetime away, belonging to another man entirely. Was it at our laboratory in Steelhouse Lane we manufactured sulphuric acid using lead condensing chambers? It warms me to recall we did, for all the dangers, make a small revolution by reducing the cost of that ubiquitous acid so easing the cost of production for a hundred things. I remember the millers happy to bleach their linen with something more efficient than sour milk.

That was another world entirely as there is now talk of a process which admits air and generates separately nitric oxide and sulphur dioxide. That I may have played a small part in the nation's 'Progress,' which quickens its pace and shouts its wares every other day, strengthens my resolve to better my farm. As I watch Watt setting his engines on the path to glory, I hope his blood quickens for inventing is the fun of the world. I am sure it has not, for one moment, crossed his mind what honours lie ahead.

*

43

'The very end of the Earth'

James Watt
Birmingham, 1777

We are thankfully home from Cornwall and feel we have returned from the very end of the Earth. My wife begged me to return, swearing she would ride pillion on a high swaying horse, or worse, walk beside me all the way to Birmingham no matter the weather or the roads and given it would probably be the death of her she would rather die happy by the road than take another week of black looks and icy winds. Had she not been laughing when she spoke I would have found an excuse to depart. I am a fool for not doing so. I delayed for yet another month of foul weather and foul workmen who were easily inflamed with drink and proud of their half-wits under the gun. Wm. God rest his soul, often predicted that Cornwall would offer great business opportunities and his perspicacity reigns true, as it does in all manner of things, but he failed to mention the hardships we would encounter and if he were still alive I would be inclined to wring his neck!

*

It was Boulton who rubbed his hands together most happily when news of our success brought Cornish miners to our door saying a fortune was to be made if we could satisfy them. I knew little of Cornwall except for the fame of their copper mines, and was most curious. We received a visit from Thomas Ennis, Richard Trevithick Senior and John Budge, (whom we had previously counted as detractors.) They had hardly bothered to wash the road off their faces and stood before us unhappy men.

We hear, said Trevithick, you are in business with the Tingtang mine at Redruth and the Chacewater mine at Wheal Busy.

Yes sir, we are, said Boulton, waving his hand in a most welcoming fashion towards our chairs.

They are dark sturdy fellows as grim as their weather and they would not sit. Instead they complained that the ease of Cornish mining has left them with a glut in copper and lowered the price of their ore as if we might be able to alter that circumstance. Budge, in particular, wanted to know if we could promise more steam power for every shovelful of coal.

Boulton, with his usual bonhomie, assured them we were eager to sell to whoever might need our engines. He described our invention of the separate condenser and our entirely new use of steam to move the power stroke of the piston. He declared our engine left Newcomen's Common Engine far behind. At this they raise their eyebrows. Being men of short speech and even shorter tempers they left unconvinced of our assurances. I told you they would be pinching their pennies Boulton said when we saw their sullen backs. Let us see what we can make of our new friends in Cornwall, and he clapped his hands most satisfied.

He was of another opinion entirely when we found a drawing of an engine was missing! It made me hot enough to declare I would go after them happy to quarrell.

We are men of business, Boulton replied and straightway wrote to Ennis most bluntly informing him 'we do not keep a school to teach fire-engine making, but profess the making of them ourselves.' [1] The drawing was promptly returned from Trevithick Snr. who likened the taking of it to a misapprehension. Boulton folded it away and laughed but I wondered what other misapprehensions these Cornishmen would put upon us.

*

The Tingtang engineer, Jonathan Hornblower is the son of Joseph Hornblower, a Staffordshire man who worked with Thomas Newcomen himself to erect engines in Cornwall. These Common Engines are marked firm on a Cornish map and, we were warned, are well-loved by mining families. I did enquire of Jonathan Hornblower, who, although upright and manly, is of senior years, if he had met Newcomen. He replied that if he had he was too young to recall for he had been forced into schooling before he took up his apprenticeship but he held Newcomen and his father as great engine men who loved God and did His work.

It took us some time to acquaint ourselves with the Hornblowers who seemed to be a never-ending tribe especially proud of their reputation and their Baptist faith. I assured him that Mister Boulton and myself readily admit we have an easier game than Newcomen who built from nothing and he nodded and raised his hands as if I spoke a Gospel Truth. I am now wary of such sentimental association, for it took me but a few minutes to judge many of the engines as clumsy and wanting adjustment at best and demolition at worse and I struggled to find words of praise for their monsters. They hiss and puff and pull their rods with such effort it is a wonder they fork (2) any water at all.

*

After months of agreeable hosts, but disagreeable servants and wretched surroundings bare of any vegetation resembling a tree, we longed for congenial society but nothing would change the suspicion put on us, where Reason itself seems sacrificed against parochial Loyalty. Fate itself seemed opposed. The owners of the vessel we had employed to bring down our Tingtang cylinder complained that the hatches of their vessel were too small! They would not alter them for pleading or purchase!

In Redruth my instructions to the engine men carried as much weight as an apprentice's suggestion. I was burdened with correcting mistakes I deemed unnecessary and the work moved at a snail's pace which left me with a pounding head and a fitful temper where I longed for dear Darwin to order me a wondrous purge to defeat my melancholy which daily grew and threatened to have me abed. Ann will have nothing of the Cornish potions but procured me nutmeg and vinegar and fussed greatly at the hardships I endured.

She was not as understanding when I took to attending the Anabaptist's services.

I do not know what you are about, sir, she said to me, but I am not inclined to be a Baptist.

Do not scold, I replied, I am in want of Solace from a Thousand troubles.

I am attending to your troubles, James, she said, most put out, and I will tell you this. I will not be made to pray on a cold floor a thousand times a day!

My Nurse turned to a Scold but the services suited me. I can hear Wm.'s quip, that Newcomen's greatest service was not to steam but to the many Souls his engines drove to his Baptist Belief!

She threatened to be worse that a Scold if I took the invitation of the younger Hornblower to go down with the tributers [3] at Tingtang to see the depth of the water myself, but I could hardly refuse for the young braggart had gathered enough of the tributers themselves to witness my reply. He could not hide his pleasure as he fussed in preparation for my descent given I was a carping Scot who had never had a day's hard work in his life! I befitted myself in the worsted stockings, trousers, flannel shirt, waistcoat sleeved and a flannel night cap and hat and took the candles and swish of gin he offered me without protest. They would have no man go down unless he was well-covered despite the heat he would encounter, that is what they told me and I did wonder if I was being led, like a stuffed turkey into a great oven. I recall I tingled and thought I might laugh I was so nerve struck!

Ann would not speak to me afterwards, and having no Wm. to confide in, I have written to Darwin to acquaint him with my adventure. I was glad I could do so as I was thankful to return with my body intact and rejoiced as quietly as I could in God's clear air.

As we climbed down Hornblower told me the mine was at 400 feet and wet, which almost caused me to laugh in his face it was such a deficit of description of a sodden hell of heat and shadows dancing like madmen. I was given a candle which I held by a piece of hardened clay and felt at once the comfort of its light. I saw men half-naked with their backs bent and not a sound from them, being no doubt, as surprised to see me as I was to be there. The boys who bring them the tools, powder and water to aid their labour were all eyes as I passed by. I told Darwin I met whole families, men and their sons in stoic spirits, working in heat beyond normal endurance climbing ladders where one slip was death. The noise of their old engine thundered through the shafts and I determined then, I would quieten any engine I made. I thought to ask about the gasses, one the called the Damp, but I was glad I could not. I could not hear a single-thing Hornblower said in the din and muck and soon indicated my desire to surface. I was then introduced to the balmaidens, the daughters of the men below. They are hardy girls whose hammers break the ore from its rocky cushions before it is sent to the stamps and are very proud of their Breton caps and aprons. They gave me the same cool look as they would a passing stranger who had no business being there.

*

Life may have been entirely intolerable had not Black sent me funds for my fossicking among mineral specimens and any curiosities I may find for his collection, which gave me purpose on my walks and a happy distraction from my travails. The land is mostly of a bluish –grey granite which has pushed upwards in great blocks which make a formidable coastline. The more I explored the greater my pleasure.

Ann and I would have a few more candles lit to spread my day's 'pickings' on our table. As the wind cavorted across the hills and the cold seeped through every crack and cranny of our rooms, we sat and marvelled at stones Black assured me will be from the beginning of Time itself. Brown and gold blistered copper rocks, sharp spotted crystals, and a beautiful smooth pebble the locals call Tumblestone. Ann so admired its black depths mixed with pink or yellow particles I started my own collection for her enjoyment.

I have been in correspondence with John Smeaton, whose reputation makes him a giant shadow over Boulton & Watt. He inquired after the possibility of fitting a separate condenser under licence from us at Chacewater (being badgered by the owners for better results) but I explained that while it would increase the depth of forking it would not save on fuel and would not prove economical for either the owners or Boulton & Watt's reputation.

The Chacewater engine is the largest of its kind in Cornwall and the Wheal Busy men are very particular about their mine, boasting its output of copper made it known as 'the richest square mile on Earth.' There were many faces waiting for me when I arrived to inspect the old giant which rises like some laird's fort against the sky. I was careful to doff my hat and asked to be shown as much as they could tell me. To my surprise they also doffed their caps and gathered around me. The mine manager gave me a bow and recited to me Smeaton's improvements which he declared had doubled the efficiency of the engine. I doubted that was the case but nevertheless I heard that the engine has a 72 ins. cylinder with a stroke of 9 ft. which worked at 9 strokes per minute.

He was especially proud of the engine's laminated beam which he confidently told me was made from twenty pieces of fir measuring 12 ins. x 6 ins, in sections and all bolted together. Smeaton had decided on fir instead of oak, perhaps because of cost, the lamination providing the strength necessary.

We measured it to 27 ft. 4 ins, in length, by 2 ft. wide and 6 ft. 2 ins, deep at its centre. It had been designed to give extra strength and I, with some concern from the workmen, insisted on climbing it. They watched with open mouths as I pulled myself up and walked the beam. When I returned the foreman told me bluntly that if the new engine can fork the water from Chacewater, it can fork anything, as that is the heaviest to fork in the whole county. Then I have come to the right place, I told them. It was an honour to work on what Smeaton had remedied, I said, for he has made us all engineers these days.

They stared at me as if they did not think it possible to reach Smeaton's summit. I told them we should be able to do better with present innovations than Smeaton was not privy to. We will get better results, I declared, because of our novel methods and our good workmanship. I said I would most likely rebuild the engine making the cylinder at 63 ins. but being superior in bore cast by Wilkinson at Broseley it will prove more effective. I said I would retain the original cylinder and use it as an outer casing for the new one. The engine was drawing water from 300 feet and I declared that we would increase that depth with ease and make great savings in coal.

There was a moment when they stared at me most uncertain like a flock about to take flight.

This great mine has worked for sixty years. Is there no man here who does not want to make Chacewater work another hundred as it deserves, I asked.

The mine manager gave me a direct look and said that if I could do that they would work night and day.

Only give me a day's work, day by day, and it will be a most certain thing, I replied.

The men muttered and clapped their hands sharp and clear and I knew they had shifted their Holy Suspicion of their 'Incomer' engineer and I may, at last, have some purchase on their intelligence; and so it was. I achieved good progress and little worry.

Word spread when we declared the engine would be ready on the morrow and I was prepared for the crowd of curious gainsayers who would arrive. I have never wanted an engine to live its measure more than this giant at Chacewater among the rubbish heaps and dank pits under the usual growling sky. And it did. There was a short moment of shock and then applause as the engine gathered velocity and thundered away making a horrible noise. Many stood in awe and seemed mightily satisfied by the din. I could see the west-country captains shuffling about and I wondered if they were counting the strokes as I was....eleven 8-feet strokes per minute; steady and strong. As the day advanced so did the engine, working with great power and 'forked' more water than ever seen in these parts. The coal boys waved to me and were quick to let it be known their engine had eaten only a half and one measure of its ration of coal. Since our Fee is proportioned on the savings in coal I was content and prayed that Boulton & Watt may see Profit at last.

In a few days, I returned to trim the engine. There was a small crowd assembled and regarded me curiously. I gave a child a penny to mind my horse which caused much muttering among them thou' I thought a penny well enough. The engine-men turned their heads as if I was about to reproof them.

Good morrow sir, said a large gentleman whom I had not met. He approached with his arm outstretched. If you are come to hear the engine you are in good time, he continued.

I thought he would shake my hand but he stopped short.

Indeed, sir, I replied. I am the engineer.

Indeed, sir, he said. I am Strickland, from the Copper Company. Are you in business with Mr Boulton?

Indeed I am sir, I replied. I am James Watt.

Then sir, he replied, the men have made their adjustments and you are in time to tell me how well our engine runs! It is a most lusty beast, is it not Mr Watt, Strickland asked, his face beaming.

To my amazement the crowd behind him applauded.

I would trim the engine to end its stroke gently, I offered him.

Good God Watt, do no such thing. I do not want a gentle engine, I want a roaring beast, and with that he waved to the crowd. He lent towards me and confided, we cannot sleep at night unless we hear its power. We are like men in love!

I did not know how to reply and so was silent. The Company are paying and the Company is mightily contented to hear its thunder echoed round the pit hills, so all within earshot will be informed that its Giant works both night and day to the Benefit of All.

I cannot abide to hear an engine used so when it can be adjusted to run smoothly and fulfil the grace of its design. I left adjustment to the engine men to please the owner and counted it as one more day before I would travel North towards the Realm of Reason. It seems to be true that an engine's noise alone gives great satisfaction to the ignorant, who seem to be no more taken with modest merit in an engine than in a man.

*

May,
Redruth, 1778

I am at Redruth to attend Tingtang at Hornblower's request. The Cornish winds seems to bring a madness, not only upon those born and bred, but even to the person of our Soho man Tom Dudley, who single-handedly and without the usual excuse of drunkenness, stifled our new engine at Tregurtha into silence! The Cornish workers have had great satisfaction in recounting Tom Dudley's *adjustments* to those about me, telling a great crowd gathered to witness the start of the engine and Dudley, (no doubt to impress many of his more churlish spectators,) set it off at a rapid 24 strokes per minute so that its thunder and fury could be heard for miles around.

I was not informed of the pressure of the crowd's applause but, as any engineer would predict, the pressure of the steam required soon ceased as all the water boiled itself away. Any engineer who has stood beside a new engine and heard it fade away to silence, would suffer a particular Mortification and I have no doubt Dudley keenly felt his mistake. It was a great credit to Birmingham workmanship that nozzles and valves and Wilkinson's cylinder, once cooled, began their work without further hesitation; its stroke set at 15 per minute.

We have further troubles. Boulton writes of a fire at Soho and a large fire no less. He begs me not to come post haste as all is in hand which is happy advice since I do not wish to look upon a burnt and smouldering engine-house and wonder how we will pay for it. He then recounts the turn down in trade, the worker unrest in Nottingham and Lancaster, and that he had reduced Soho's workers by several hundred! He writes that Fothergill is dolorous to the extreme and thinks only of bills and wages wanting payment. Fothergill is all for bankruptcy but Boulton will have none of it. He has, instead, and in defiance of caution, borrowed £17,000 against our patent!

I do not know where madness is more located, in the wilds of Cornwall, or the Manufactory of Soho! Boulton writes I must obtain payments on the mere promise of building engines! If I was not averse to a surfeit of madness, I would advise him to come and talk to the Cornish Mine Captains himself!

The Midland bankers may splash money like water but a Cornishman would rather die of thirst than be loose with his money. But I have heard there are two thousand men at Wheal Virgin who will have no work if they do not clear their water.

*

These Cornish engineers go from one extreme of the compass to the other. John Budge is our convert and parades his praise of Boulton & Watt like Paul on the road to Damascus while Trevithick from Wheal Union puffed out his chest at me and told me he thought our boasted savings in coal a great romance. He had his nose an inch from mine as he spoke and if I had not thought of Boulton laughing at their childish stealing I would have created a new reputation for Boulton & Watt which would have set tongues wagging and would have had me in conference with a magistrate.

*

44

'Watt suffers....Boulton runs full tilt'

Erasmus Darwin
Lichfield, 1778

Were it not so painful I would take some pleasure in refuting the great Aristotle for I declare I am, at the same time, both one thing and another. I no longer spend my leisure hours in pursuit of Truth by Reason Alone, as we Lunarticks declare is our life's blood, but instead walk in my garden to weep away what is left of the day or sit at my desk bewildered by love which has seen fit to strike me well past the noon of life.

Both Grief and Love are great Bullies who push and pull and poke at me, agreeing on only one thing, that life is not to be borne without the missing loved one. My dear Charles, my first born, is dead, with nineteen years only on this earth. Yet I carry a traitor's heart of leaden sorrow, which, on an instant, can melt to warm tissue for Elizabeth Pole, whose eyes are only for me. It is of no consequence to my heart she is already taken by a soldier husband given to the happy dominance and hunting skills so favoured by soldiers and ill-favoured by myself.

Despite every remedy known to Doctor Withering[1] and myself, Charles succumbed to in infection caught in the autopsy of a dear child he had treated for Hydrocephalus. His lecturer told me he went quickly from headaches to putrid fever and when we examined him we found a wound, slight but weeping still, as the open door for infection.

And thus we lost him, and he lost the world. Such brilliance in medicine that lay ahead never to be. And alive within my heart there is Elizabeth Pole, the daughter of an Earl, the most darling, witty, dark-haired beauty. She is a mother of three and in her prime and fills my dreams with such passion I wake hard and sweating. I cannot rise before such release as is healthy for all men.

I have turned to Poesy whose form allows a man to express the grandest of emotions and steady his soul while all about me is turmoil. My dear American friends have taken up the gun against the King and I fear not only for their lives but for the perfect principles of Liberty & Equality they profess. Boulton, being able to mesmerise bankers like no other man I know, borrows more, and mortgages more, month by month. Others would seek to cut their cloth but Boulton has the courage of a lion to advance whatever is his heart's desire. Meanwhile Watt fears a return to Cornwall where he was at the mercy of his headaches, the terrible rain, drunken workmen and fear of the Poor House. Boulton & Watt have taken the world by storm as we Lunarticks knew they would, but Watt suffers anxiety at what lies ahead even as Boulton runs full tilt to embrace it!

*

Aris's Birmingham Gazette
April 20, 1778.

'The following Letter received last Week by the Committee of the Birmingham Canal Navigation, from their Superintendent of the Locks, affords an irrefragable Proof of the great Utility of a new-invented Steam Engine, lately erected on the said Canal, under the immediate Direction of Messrs. Boulton and Watt, the Patentees.

To the Committee of the Birmingham Canal.
Smethwick Locks, April 17.

Gentlemen, - On Wednesday last Mr Smeaton made an accurate Trial of the Steam Engine erected lately on the Canal at this place, and it appeared that it did not consume more than 64lb of Coal an Hour when working at the rate of 11 Strokes a Minute (each Stroke being Five Feet Ten Inches). The Diameter of the working barrel of the Pump is 20 Inches; and the perpendicular Height of the Column of Water is 26 Feet 10 Inches and a Half, equal to 11-lb. 3-qrs. upon every square Inch of the Piston: The Quantity of Water raised at each Stroke is equal to 12 3 qrs. cubic Feet. Mr Smeaton declared, that the best new common Engine, with all his late Improvements (which are very considerable) would have required 194 lb of Coal to raise an equal Quantity of Water to the same height; and that a common Engine without those Improvements would consume a still greater Quantity. "When that Asperities on the different working parts of this Engine are worn off, and the Cylinder is eased and finished, as is intended, I have not a doubt but it will be an Advantage to the Proprietors of 20 per cent more." "I am, Gentlemen, your most humble Servant, S BULL.*

Original Document. Aris's *Birmingham Gazette* April 20, 1778. John Smeaton's measurement of the effectiveness of James Watt's new steam engine.

45

'as bold as a page from Revelations'

Andrew Cosgrove is 49 years old and a Master Blacksmith.

Andrew Cosgrove
Handsworth, 1779

I am an odds-bodkin of a blacksmith. Words, which have long seemed strangers of no necessary purpose, now pile themselves upon my mind as if they are so many glowing nails to be bent into shape befitting the hammer of some grand design. Whole sentences come to me with the ease of bending a shoe white hot from the morning bellows. As my dear wife has often foretold, a man who can read, and does so as often as I do, should be no stranger to a line of words.

My mother was full of stories when she paused to regard one of her children and often bent my ungrateful ear to some matter small in my reckoning which had delighted her, but was for me a thief of my time. She would smile that I take up this telling, even though when the ink dries it may be no more than the worrisome laments of an old man over his ale. But whether a story proper or a singular lament, the naming does not still the strange wonderment and fear which fill me as I write, having witnessed the future as bold as a page from *Revelations*.

I should start my story from the day I heard a rolling Prophecy like so much thunder from a clear sky and had no need to look up because I saw it all before me and knew whence it came.

My wife's brother being an Apothecary knows Doctor Darwin, and it was the good doctor who recommended my services to James Watt, the Scotsman who came into the district on the coattails of Mr. Boulton of Soho, and who lives on Harpers Hill at the top of James Street on the edge of town.

I had been called to shoe his little mare, which I was happily doing till Watt himself came down the hill which runs at the back of the house rubbing his hands together in such excitement I would not have been surprised if he had found the goose which lay the golden egg among his chickens.

But no, for he asked, Smithy, my man, if you could spare me some time and beckoned me to follow him. I happily complied and he led me up the hill to his Office as he proudly called it, being indeed a separate but small room at the side of the house. He seized upon my presence to exclaim the beauty of his latest "invention" a heat driven miraculously by steam propelled through the whole room to make his office comfortable through winter.

Not content with my astonishment he begged me to sit while he explained his new measurement of "horsepower" presuming by my trade that I would join him in swearing the truth of his calculations. Indeed, I was drawn to consider his claims; having lived so close to these wonderful beasts over many years, I would place a coin or two on my judgment of the number and size needed for any industry a man might ask of them.

Watt showed me his notes where he had written that a horse in general health can raise 150 pounds by nearly 4 feet in one second, making it 550 foot-pounds per second. It would be a strong horse I thought but perhaps not so far from the truth of it that I would dispute a capable gentleman.

But he turned to quiz me directly as if my reckoning mattered to his thinking.

What other power do we have but the mighty horse to measure the rate at which work can be done? He waited for my answer his head cocked to one side and his eyes bright like some blackbird waiting at my cottage door.

In general health, I enquired for I had heard he was most interested in the working strength of pit ponies.

In general health, he replied.

And happily disposed to his task?

Indeed, Smithy Cosgrove he said, and smiled most kindly at my concern for his fantastical work horse.

A healthy horse kindly disposed to his task would be stronger than a pit pony I ventured.

Indeed, that is likely true. I have measured a pony to lift two hundred and twenty pounds to one hundred feet in a minute. I have added to the weight and distance, he replied, flourishing a quill pen so hard-used I doubted it could make nothing more than a blot upon any page.

They are a hardy breed I replied, amazed at our conversation and much taken up in it. They are much loved and will get a miner back to safety in the dark.

Watt paused and nodded.

They are mightily useful, he exclaimed. I have come to the calculation of thirty-five thousand, five hundred and seventy-two foot-pounds of work every minute.

He folded his arms across his chest and seemed very content by this estimation.

Then Mister Watt, sir, it seems you have made a singular Calculation, I said.

He smiled, most satisfied, and I do believe he thought it sealed until Kingdom Come. I made my excuses to finish shoeing his lovely little mare that was inclined to kick, being far too lively from under-exercise as Watt buries himself in his Devices. It was as I walked down his hill I heard the rolling Prophecy as distant thunder and turned back to the outline of Watt's Office against the blue sky of Heaven with a heavy heart. There were no dark clouds but thunder trembled my bones.

Absalom was waiting where I had tethered him and I paused to rub my hand across the velvet of his nose which pleased him greatly for he thought I had mistaken the time and might offer him his half apple. But all I had was a twisting in my guts as I stood beside my warmblood who towers over me. He has in him a touch of the Lincolnshire Blacks and is the best horse in Staffordshire as I will tell any man who asks.

I rode home like a man in a dream for it has settled in my bones that Watt had determined his measurement for another purpose entirely and that his "steam" power will be measured to outshine the horse with all his Majesty and Strength, and so render my trade and many others, mere tinkering. Mistress Cosgrove has taken all my protestations with a great silence and finally pronounced my fears as mere Chimeras and Fancies of my digestion because the horse, being so plentiful, easily fed and loved by all, will triumph over such whimsys by which Watt thinks he can seduce the world.

*

I write under compulsion.

My wife has sat down with me over my lunch at the forge which she never does the forge being far too hot and dirty for her clean aprons, but she had in mind to speak and so she did, and my ears are ringing still.

I married a blacksmith not a dull-swift, she began, who is confused enough not to know a goose from a gander. You must be at the mercy of some devil who has put a black cloud in your noggin and I am about to shake it out. Counting Sunday to Sunday there has been a week of you staring at the sky and I will have none of your harecop nonsense. Tonight I have a quill for you and you will exercise your thoughts to remove the demons within for I will not have you brooding by the hearth cluttering my busyness, for all manner of tasks, she assured me, must be done before Doomsday is upon us. I half-expect demons to delight themselves dancing across my page but I did as she directed nonetheless.

I have noted that many travelers read books seemingly of a comfort as they wait huddled near my bellows in all weather, so it is possible that my words may one day dwell upon the thoughts of others, for there is a Monster to be set upon us all and it needs telling to those who may know to listen. If my thoughts be topsy-turvy that is the nature of the Monster for it is a marvel to behold and sets me at odds with myself.

Yesterday Watt sent a note asking me to bring his mare early to Smethwick, where he is currently working, so he could go about further business if he needed. He has taken a liking to me and I am happy to oblige since he sends Business my way. I found him west of Bridge Street on a hill above the first lock in the main canal with a great working party set upon the erection of his engine for *Raising Water by Fire*. My wife would have thought him foolish not to have a cloak against the morning cold but he was deep in meditation with a mason, bent forward as if he was set against an invisible wind. Having caught his attention, I left the mare tethered nearby and had every intention of getting back to my Bellows forthwith but the clamour of the whole Proceedings compelled me to pause a while, if only to understand what such a collection of workers hoped to achieve.

Smethwick village has not seen such a commotion since James Brindley's men built their canal. There were many bystanders whose Curiosity was greater than their need for work that morning, including their wives and children. Men stood measuring the height of its engine house with their arms outraised and their hands forming a point, and then turned away shaking their heads. It was clear to see the foundations being laid, such as could be used by a needy farmer's family. I heard one farmer claim that the boiler could be 20 feet by 12 feet to fill the cavern left for it, and nothing good could come of such extravagance. If he had not left in a storm of feeling I would have assured him that such a boiler could not be made.

I stood near masons stamping their feet against the chill of the spring morning, the sun feeble above the nearby fields and the grass waiting to feel the lambs a dancing. Since I am known by many as the Smithy of Handsworth, I was able by friendly interviews to gather more news of the activity around me. The masons happily boasted of their contract for they were to build the boiler pit and engine house all double walled and plumbed true as a permanent scaffold to cover the whole enormous engine and its tunnels to the canal, which would take near to two hundred thousand bricks in their estimation! Watt would have many manufacturers happily do their sums! I could not deny, despite my Premonitions, the audacity of Watt's plan makes me tingle and want to slap my thighs.

Every man was intent upon his task and I wished I'd kept my apron so I could have merged like a salamander into the site. I stood near the pits, which seemed by my reckoning to be over 12 feet into the ground, and watched what I took to be the steam cylinder chamber, because it was deep and round, being prepared by masons. They wrestled to lay three massive sandstone blocks on iron plates. They were in differing hues and about them lay long iron bolts squared at the ends as if for retaining keys. I could only surmise that the colours signalled the hardness of the stone and its mineral content but as for the harness of iron I was at a loss.

By then the morning had blossomed into a fine day and the sun warmed its way through my coat, and it should have caused me pleasure for once not to be bending my back, but I have never felt so much a bumpkin out of his time and place for having the temerity to seek to understand workings well above my station.

The engine house presented its arches and walkways and tunnels as if it were a secret lair for some huge work horse needing a stable. As I looked over the whole works and took in the rise of the ground to the summit the better to acquaint myself with Watt's ambition, a wild laugh rose in my chest as if I had heard a rakish joke at a Publick House. I had my Kerchief to my mouth in great haste lest I offend those nearby. There were idle hands gawking about but none seem to have estimated the task which might give Hercules himself pause. Surely no pump could lift water back up the canal summit locks, a rise over four hundred feet? I stood transfixed as if caught in Watt's wild fantasy, uncertain as to the sanity of my fellows who put their faith in a wet, hissing Vapour that vanished in a twinkling.

There are few who have not seen steam make merry with the lid of a Pot and I have seen it scald a man's skin in a flash and fester for many a day but I cannot see from whence its power comes. Besides, any man surveying the lie of this land (1) would recommend a tunnel be dug so the locks at the summit were not needed making a nonsense of this fuss around me. Yet I found no one who would gainsay Watt's enterprise and several boasted as if they knew him personally, that this was not the first pumping engine he had provided and that a small steam engine was raising water at the Spon Lane top lock and was already causing boat captains to praise it.

There was talk of the engine's beam, by that I assume the giant wooden arm which moves the pump, was to be held by its brothers and young fine-grained oak pillars have been ordered for its support. The iron struts, braces, bolts and chains are all to be cast and shaped from the forge in the hands of the ironmongers from nearby Coalbrook Dale. It was no surprise to hear of Coalbrook Dale because in Handsworth we do not need to go far to know of these Iron Makers, and travelers speak of furnaces that belie the setting sun and burn golden fire clouds through the night to defy the laws of God Himself, as my dear wife declares.

I have never had cause to go to Broseley or those parts where Darby and Wilkinson have their foundries, but whatever admonishments fall upon them from the Cosgrove Hearth, they cast the finest iron that any of us Smiths have seen. Further, the shire is agog with talk of Darby's Great Iron Bridge being made across the Severn Gorge, an amazing feat across that vexatious waterway if Darby's men can do it, and threaten us with another Miracle at the hands of Man to wonder at.

Further still, it is Wilkinson's iron which is to hold Watt's awful engine to account for itself. Wilkinson is a man who sets great store on his iron for its strength and endurance and many were singing its praises, which I put down to a bit of tilly tally since none of them had ever lifted an anvil hammer in their lives. For them the whole assembly was a wondrous giant in the making and I watched a long line of contractors delivering their wares through the morning, rubbing their hands with excitement to see what may come of it all. I wondered if I had become that fool who seeks to find fault with a fat goose!

I was making my way back to my own horse when a Team of Four arrived and since I knew them and their driver, I could not keep my Manners and not make myself known to him and enquired as to the health of his family and if his business went well. He had on his dray a long iron cylinder just come from Wilkinson's at New Willey which was as heavy as the Devil's night pot, he told me proudly. His horses seemed calm enough and welcomed my touch and some leftover apple but the driver was wide-eyed. He pulled back the covers and placed his hand on the side of the great cylinder, remarking that he had never seen the like in all his days. I reckoned it was at least 12 feet long and 8 feet in diameter, and could not believe my eyes until I made out that it had been cast in separate lots and joined by driven rivets.

I do not know who might have observed us, but we stood like green apprentices, both of us, with our hand upon its hard surface, our eyes narrow and our faces drawn as if by touch we could understand its Provenance and lessen its Magic over us.

It is cast you know, the driver remarked as if we had been discussing its particulars for hours. Cast and made from the ore mixed with coke in the Darby way, he added.

And he has made a wondrous boring tool besides has Iron mad Wilkinson, he continued, his voice low as if he was parting with a King's Secret.

This cylinder is true to the edge of a shilling piece, he whispered.

I nodded as if this was nothing new to me, but since my hand had never touched a man-made surface of such Width and Might, I had nothing of worth to add. It seemed I touched a marble rock whose skin spoke only of its unchangeable Strength and was made to last until Kingdom Come.

The whole district knew of Watt's promise to the Navigation Company of more wealth than they could imagine, which would need to be the case since the whole district also knew that the cost of the Fire Engine was said to be many thousands of pounds because of Watt's insistence upon the Singularity of his invention and the protection of his Patent.

Would it pump water back up the locks at the speed and volume Watt had promised? I knew I would find some pretext to return, for I dearly wanted to learn if Watt's engine could lift that weight of water with a power beyond any team of horses known to me.

*

46

*'contemplated him as the rat in the barrel of
my future and could not hate him for it.'*

Andrew Cosgrove
Handsworth, 1779

It has been a long day made of many parts which need time in the telling. I must report that the summer has been kind to us, the weather warm and my business prospering. We have Josiah Winterson with us now as Tomas, my apprentice, asked me to take on his brother which I did at Amy's request.

It has taken me many years to learn the full extent of Amy Newton's family which seems to have been blessed with continual Growth. In truth, I have married the Good Samaritan's sister, who lived undiscovered in Leicester until I asked for her hand. The Wintersons, being close to one of her numerous cousins, were in need, for they are weavers and their trade has fallen on hard times and they have sent their girls into Arkwright's Mills. Their boys are a long way from home at Handsworth but they are hardy and if they were myne I would rather have them blacksmiths than at the beck and call of the Spinning Jennys crammed, as I have heard, hundreds to a room.

I have had much to do as many housewives are in need of my skills. If I were to say this my wife would lift up her eyes and roar with laughter and teaze me till she would say I blushed like a cock at dawn. We have each other's measure, my wife and I, and like a well-balanced Maypole we often take to dancing.

The repair or replacement of kitchen items and Carpenters keen to order my clean Nails and have their workhorses reshod have meant a merry crowd round my Bellows and gossip aplenty about the news which takes their fancy. There is nothing I hear more of than the Fire Engine at Smethwick. It is ready to start on the Morrow and many were finding a way to have their Business in Bridge Street or on the canals, the whole district abuzz with news of its function at last.

Late in the afternoon I had a note from Watt requesting my attendance at a time that was most inconvenient for me. I was heartily annoyed at this, for it assumed I was at his Beck and Call like one of his workmen dithering round his engine. Tomas, the older Winterson boy, stood from his rubbing work and came to me directly.

Master Smithy, he said, giving me a happy smile, leave it to us to finish up the day, for besides, Mistress Cosgrove will be here soon with our cake and we can close the Manufactory together.

Cake, is it, I said, with a look of Horror on my face.

I had to stifle a laugh at his flattery, for my workshop could hardly be called a Manufactory. It was a very fashionable word and it gave him great pleasure, but he was well off the mark to reckon Mistress Cosgrove would need any help to open or close anything.

I told Tom Mistress Cosgrove had been my Foreman for many years but it was his Duty to take care of his brother Josiah and that he was not to use the bellows on the Forge fire, but only to complete the task I had set him. He nodded in delight and tugged his forelock with great respect to his Master but I was glad that Amy would be here presently to oversee them both.

With that I rode into Harpers Hill at a brisk pace lest the day escape me. I found Watt out in his yard, contemplating his mare, with a cloak wrapped tightly around him and a wan look on his face.

Ah, Smithy Cosgrove, he said, looking so mournful I could have been an Undertaker come for a body rather than a Blacksmith come for a job, thank you for coming. I am in need of you.

He did not seem in a good way and neither did his little mare that I observed was limping and had thrown a shoe on her front hoof, for the second time in my reckoning.

She will not let me near her he explained his voice thin within the folds of his cloak.

I muttered a reassuring grumble for I was at a loss at finding him so reduced. It did not take me long to run my hands over the mare and sooth her offering a sweet apple core to show my appreciation of her Predicament, which was easy to diagnose. She had managed to throw the shoe but left a nail now bent towards the tender part of her foot. I would be limping too I thought if I had to bear my weight on such a bruise as it was making. In no time at all I had removed the nail, replaced the shoe and rubbed a salve of my own making into the wound.

Watt would not let me leave but beckoned me to his work shed, offered me a comfortable chair and asked if I was of a mind at the end of a long day to indulge him, would I share a jug of Cider? I was not one for Cider but not wanting to disappoint him I thanked him for the offer and added I would be glad to partake as long as I personally knew the rat that had been left in the fermenting barrel.

He was most surprised at such a suggestion, and was amazed when I told him that local Cider makers threw a rat into their barrels. I added I did not know the reason for the custom but assumed it was to add Flavour of some sort and pulled a face at the thought.

Perhaps, he said, looking thoughtfully at the jug in his hand, but I consider it is more likely a way of adding minerals to the fermenting process.

I nodded in appreciation being suddenly convinced he was right. I became alert in a way I was with no man I knew, for in truth Watt was like no other man I knew. He unsettled me and inspired me at one and the same time, for I was often troubled yet uplifted in his company as if I was in the hands of a Magician I had not recognised among my Acquaintances.

When we settled back to enjoy our Cider I noted he was again wan and sought his cloak in case he had a fit of the chills, for he did indeed look like a man on the edge of a Malaise.

You have heard that tomorrow is the Canal Company's Lofty Occasion, he ventured at last.

Yes sir, I replied, the news is all over the district.

Well, he said, it is always a great impost upon me. It is never easy the night before one of my Engines begins its purpose. He spoke as a man whose audience might damn him with hissings and boos if he dashed their grand Expectations.

I felt a chill also. The Cider was like honey on this summer evening but it did not warm me. I had heard the giant Engine now built at Smethwick was one of many. I had a flash of them spreading like some frost plague across the fields!

I have heard this is not your first such Engine, I inquired.

O God no, Smithy Cosgrove, he replied, his voice gaining some interest in life. This is the fourth such Pumping Engine I have devised and Mister Boulton and myself have put our names to.

I listened, amazed to learn of a world beyond me, a world so unknown to me I could barely imagine its form.

I have one working near Tipton, he explained, pumping at the Bloomfield Colliery.

Well, sir, I am a Smithy well behind the news of the times.

Watt laughed. There is an Engine at Wilkinson's ironworks at Willey which you must have heard of being associated with your trade.

No, I have not, I said much puzzled for the news from Willey was entirely about the coking coal and the quality of the new iron produced.

Well Mister Watt, sir, you are a remarkable man.

He looked at me directly.

And how would that be, he asked.

Because your Fire-Engines will change the course of the World, I replied. I could not return his gaze as I was so taken with the import of my words.

No, no, my man, Watt continued. The history of the World is a vast rolling ocean and I am but a small raft upon its surface. I cannot make such a difference as you are imagining.

I could speak no more, deep as I was in the profound nature of our conversation, and feeling well out of the depth of water Smiths wash in. I sat back in my velvet chair and drained the last of Watt's golden Cider and contemplated him as the rat in the barrel of my future and could not hate him for it.

*

My legs ached to rise and I would have taken my leave but myne host continued to chat, seemingly in need of company, though why my company in particular puzzled me, for I was not one to waive my bill on the strength of his Cider.

He kindly inquired as to the strength and knowledge of my horse for my safe passage home. I assured him that I owned but one horse, Absalom, and he was the best in the County, to which Watt smiled and said he had no doubt of that.

I drank a passable amount of the Cider and gnawed like an anxious mouse at his cheese. My wife would have called me a fool not to eat more, but I was aware of the passing hours and her imaginings as to my whereabouts.

Watt spoke of the men he relied upon and whom he admired, and promised to introduce me if there was an occasion to do so. He complained of the constant threat of competition from rivals who would challenge his Patent. I knew very little of Patents but it seemed to me that any man who publickly bricks in his treasure should expect thieves to climb his walls.

You may know John Smeaton, a man of great reputation, he asked.

I did know of John Smeaton through Amy's cousin in Smethwick, who had come down from Leeds in Yorkshire to marry a Distiller and often spoke of the great engineer from her city as if he was rebuilding the Roman Empire from wild forests.

He has been a doubter of mine and is called upon to make us answer to the loch owners, Watt said.

He is a Yorkshireman, I was bold enough to comment, and he will call a spade close enough to a shovel for the sake of an argument.

Watt gave a short laugh.

He has been forced to reconsider, he continued. He has given a favourable report to the Canal Committee on my Spon Lane engine. He estimates a saving of two-thirds in coal.

That is a considerable saving, was all I could say.

It is, said Watt. But every engine must keep to that mark. And it is one of your iron men, Cosgrove, who aids our success.

How so, I inquired.

John Wilkinson has devised a tool that can bore a useful cylinder.

And how did he do that I asked, fascinated, for I had seen such a cylinder and laid my hand upon it.

Watt's face took on a glow and his eyes shone in enjoyment of his subject.

Wilkinson realised stability leads to accuracy, he explained. His machine is supported by a shaft that holds the cutting tool at both ends and he devised that this should extend through the cylinder.

His hands moved through the air as he described the tool and he smiled as a proud father might when Wilkinson, from my knowledge, was well his senior.

He has given us greater precision, he continued, and that is most what I desire. He began to fidget and I thought it well past time I should take my leave of him. He stood and brushed himself down. I thought he was about to show me to the door and I stood also.

They are all different, he said quite suddenly his cider mug still in his hand.

How so, I replied for I knew he spoke of his engines.

I wonder if it is not only my adjustments and improvements but the lie of the land they are on, or the shadow of the sun, or some such other thing, that a wise farmer may know of his animals but I do not of my engines.

You have given it a very solid home, I replied, it should not want for shelter. Watt gave a small smile.

Thank you, Smithy, he said. The engine house covers the whole engine not merely the steam end and it is a veritable rabbit warren of tunnels to every corner I could devise.

It has certainly made the brick-makers happy men, I said, feeling that Watt was about to preach on his Engine House and I should be gone before he did.

Yes, indeed, he continued. I have added a tunnel so a man can inspect the cylinder for any damage caused by vibration. And I have estimated this Smethwick engine should raise 230 gallons with each stroke.

If I had not been a fool I would have let his words pass by as hopeful Fancy but I was so amazed at his estimates I asked how many strokes a minute he foresaw.

He raised his head quickly as if challenged and colour flowed into his face. I am in expectation of 15 strokes a minute, he replied.

I could barely grasp the calculations before me but I exclaimed that 230 gallons was some 1800 pints at 15 strokes would resemble many thousands of pints. Perhaps it was the cider but my brain was a blunt as a bent nail.

Twenty-seven thousand pints, Smithy Cosgrove, Watt said most calmly, we must refill the canal at the top of the locks in a timely manner for we have said we will have two hundred and fifty boats a week pass through the locks.

I did not comment on the two hundred and fifty boats for it seemed beyond possibility. As we stood together I thought of a man called Sisyphus in a story my wife once told me. This man was made to push a stone up a hill and just when it reached the top, it rolled down again and, over and over, he had to recommence his struggle. Watt was like this man, and I wondered why the Lord had so burdened him.

I went to speak but he looked like he might fall asleep where he stood, like Absalom with his face to a breeze. I took my leave and he hardly heard me go. My wife was anxiously waiting by the hearth that she had built to a high old blaze, expecting me for some reason to have likely fallen in the Severn, which was miles away. Being unaccustomed to my Experience I hugged her soundly and took comfort from the warmth of her breast against mine.

Come to bed, Mistress Cosgrove, I said, I promise to tell you everything.

*

47

'With the utmost rigor I have observed Watt's engine'

John Smeaton
Austhorpe
West Yorkshire, 1779

I have written to Watt, although I have held off these past months and given much thought to his achievements which grow, seemingly every other week. My report to the Committee of the Birmingham Canal Navigation Company, I have no doubt, will be but the first of many to assess Watt's engines as the most remarkable advance in steam power yet made by any engineer here, or on the Continent.

Watt has not tinkered with steam power, tightening here and there, as I have done, but used Science as She should be used. His separate Condenser and use of latent heat is based on the most profound understanding of physics and is to be greatly commended.

With the utmost rigor I observed Watt's engine at Spon Loch. The Common Engine cannot win in any comparison known to me. To raise the same amount of water Watt's engine used 64lb of Coal which compares most advantageously to the 194lb of Coal I calculate would be used by the Common Engine, a saving of two-thirds as Watt has predicted. Watt's engine produced 113/4 pounds per square inch of the piston but is, I predict, a pressure shortly to be bettered by his advances.

I do not doubt, for one moment, Watt has only begun his Improvements, and, were he to desire as much, I would happily give testament to whatever he may devise.

*

48

'I heard what I heard'

Andrew Cosgrove
Handsworth, 1779

We rose early and my wife made no comment as I prepared my horse to ride to Smethwick. As I mounted she came towards me. I would accompany you, she said, but one of us must attend to business. She said it not as a rebuke but as a reassurance.

I will be here she said, wrapping her shawl around her, and stroking Absalom with calm hands. I will be here as will be Autumn, Winter and Spring, she might have added, as she reached up and touched my hand.

I turned towards Smethwick and thanked the Lord for Amy, but I did not thank him for what lay ahead on this Day of Days for the Birmingham Canal Company, Boulton & Watt and the Ironmen of the Severn Valley. The June air was fresh and gorse coloured the hedgerows but as I rode I could think only of my predicament. I felt akin to a condemned man who eyes the scaffold upon which he is to hang merely to judge whether it would hold his weight and finding himself hoping that it would in admiration of the carpenters. There was a steady stream of travelers on the road, all sorts looking spick and span with not a crease about them and jaunty with it. I was one of many bound to see the great 'Fire Engine' we were told would lift a summit of water and amaze us all.

When I arrived, the crowd was considerable and I caught a glimmer of Watt in the distance, his head bent in conversation, surrounded by the Notables of the Company. I could not have imagined I would have seen so many powdered periwigs all at a time, and so many faces flushed with rouge. I saw the masons and smiths turn away to hide their smiles and I could not help but do the same. But the men who gathered round

Watt were serious men of business, for all their puffery, and they fiddled with their canes and waited barely patient for the Firemen to be finished with their lighting and stoking.

It seemed a long wait too for those of us stood in the dust by the edge of the works at the mercy of the noisy musicians sent to make an Occasion out of what was, for Smethwick, the grandest, most unusual Happening ever witnessed in the district.

I heard the first sound as a soft thud, and then another as a muffled beat of a drum. Not one man turned to his neighbour to remark upon the sound but slowly we fell silent, listening till the slow thud became a booming regular drumming as an oncoming army of Giants might make, and from the front ranks of on-lookers and spreading back towards us came a full-throated cheer and hats were thrown in the air. The crowd surged forwards for it heard what it had come to hear, the full gush of water moving with the giant pump as if it had the force of a gale behind it. Water is a heavy weight to move and every Boat Captain watching knew what a difference a speedy pump would make to their passage through the Locks.

I could see some of the Company men had their pocket watches out and I caught a glimpse of Watt, leaning forwards, as if he was listening to the Giant's heartbeat, and the same intent and anxious expression on his face as he had had the night before. I would be shouting Halleluiah with relief, I thought, for the Beam moved smoothly and with such a confident drumming, nothing in this world could be its equal. But I remembered enough to know Watt was counting the strokes per minute and listening for the hiss and wheeze of any air escaping his cylinder and that every minute would have been a test of his Mettle.

I had seen the Engine commence its work and stood in wonder with the crowd. It was then I should have taken my leave of the various men I had met as they toiled to bring this Giant to life. Instead I took leave of my good sense and pushed forward. The Company men with Watt had moved to inspect the water flowing into the top Locks and only a few attendants stood with the Engine. The engine house gleamed in the morning sun, its new bricks as if polished. The ground did not shift or tremble such was the care taken with the foundations, but the air was heavy with the smell of heat on metal, a smell so familiar to me I was taken by surprise as if my Forge was the only place on Earth for such a smell. It drew me closer, being so warm and rich and familiar. It was not a cold morning but I felt in need of warmth and stepped closer still. I wanted to call this Engine Friend, and lay my hands on its sides and feel its pulse through the pumping rod rising and falling but I dare not go closer.

I could but stand and listen as many men about me were, for in truth, none of us had heard such a sound in our lives. It was an animal of iron and brass, most content to lift its iron trunk and sigh a deep wet sigh, as if it needed to please no one but itself.

It was not long before I turned away, soothing Absalom who, even at a distance, was not happy with the noise, and rode back to Handsworth and Mistress Cosgrove post haste.

*

My wife had the boys lunching with her and pointed to a fresh loaf, a side of beef and good ale she knew I would enjoy. I might have been a soldier returning from a battle she took such care of me which made me feel weak at the knees for I had something of great moment to share with her.

When she had sent the boys about their tasks she looked at me expectantly and took a hearty drink of ale herself.

Well, she asked, and put her head to one side like a bright-eyed bird.

She looked so intent upon my happiness I held back my thoughts.

It is a fine Engine I said, pumping very well, and a great success for Boulton & Watt.

She clapped her hands together. I am happy for the Canal Captains she said, rising to gather our eating plates. It is good to be done with that business, she added, giving me a smile and placing my work apron in my hands.

I went back to the forge and began my work but as I turned an iron post and had Tomas striking, I heard what I had heard when I stood close to the Engine and wanted to call it Friend. It came from within and I heard it again as if my ear could hear the singing of an underground river or a Winde high above the Globe swirling through the Heavens!

It was clear and it said - I am the Alpha and Omega, it said. It paused and begun again as if I had not heard. I am the Alpha and Omega, and I heard clearly. I am the Beginning and the End, it said, smooth and steady. The Beginning and The End. It was the voice of Prophecy broken from a page of *Revelations* I had heard before at Harpers Hill and I knew it to be true. I could explain to any man in a voice of Reason that Watt's engine marked a beginning because it was an Invention to mark the Age, with its Progress and Perfection, lauded by all who speak of it - as to the rest I trembled to think. I had long feared it would outdo the horse and now I feared it would grow to outdo any man who is the maker of things.

I bent to the rhythm of Tom's strike and the simple turning I had mastered for many years to make all manner of things to lighten the load of working people. This is what I knew and I sought comfort in it. Who was I to say I had heard such a thing and try to fathom its meaning? I am no man of learning and my wife would say no good had come of my association with Watt to lead me to such Delusions. I had called him Sisyphus with his rock but he was not a man in Punishment, he was a man driven by Struggle, and that was where he was Happiest and he would continue until the Lord called for him.

Today I looked upon the earnest face of Tomas and swore I would make of him and his brother such Smithies as the whole district would ask for them and them only, and they would live in great prosperity with their families, and Mistress Cosgrove and I would live a long age to see them so established.

*

49

'I am as busy as a bee in a bottle'

James Watt
Birmingham, 1780

I am as busy as a bee in a bottle which would make me *all abuzz* as my Jim would say, dear boy that he is, a gangling lad at eleven and hopefully improving at Winson Green school. Boulton has become a regular will-o-the-wisp with his time. He writes he has much to take his attention and has left me to ache and toil as if his signature does not appear on any contract given to Boulton & Watt. There was once a time when he would deliver new lead pencils upon request so I might be free to ponder my latest "sensation" as he called it. He has not thought it once worthy to rise at dawn to review what his engineer is up to freezing in the first light of God's morning. If it were not for our letters and the most reliable post boys he may as well be in a cave at Land's End for all I would know of him. He has left it entirely up to me to oversee the multitude of adjustments and engine orders and leaves me his Soho man, John Harrison, by way of apology. He is much regarded by Boulton, who had him reassemble the Kinneil engine, and shares the same carefree optimism for the happy outcome of every endeavour. He may deliver all valves and the pistons complete to specifications of values, but he is a rough and ready man who worries me with his nonchalance. Darwin is kindness itself but he is so enamoured of Boulton, I cannot find in him an ally.

Meanwhile I have grown a thousand aches thinking of any return to Cornwall and my wife Ann hounds me to wear a scarf and cape against the canal damp. I bend over my plans like an old wizard with an aching head, searching for new gains with which to impress the Navigation Board.

There is a grand idea that has been on the mind of many but plagues me daily. Darwin and Boulton will not let me forget their fiery chariots and scribble in their notebooks various fantastical designs. John Robison writes from Edinburgh University that he expects a steamy wheel any day now and reminds me it is a while since we were young conspirators in Glasgow. Joseph Black talks always of my Advancements as if he expects me to announce an engine with rotary motion any tick of his pocket watch. For that is what fills my mind like Ben Bold waiting to be climbed. It seems all manner of men talk of the 'Necessity' that will conceive steam driven wheels if only Boulton & Watt would put their mind to it!

I have a way forward that I have begun, but they are merely scratches on paper. Boulton has left me to bustle and bargain for contracts. I have long hated the whiffle-whaffle of commerce and would rather face field artillery than blunder about, as I do, making a bargain. For days I am quite eat up with the mulligrubs, and to complete the matter I am often obliged to go to an oratorio, or serenata, when my mind begs for oblivion.

Ann keeps busy days at Regents Place with the children who are sickly and in want of her care, and I keep a lonely workman's life. Thankfully she is content with the house, it being a substantial construction built of brick with stone facings and in the happy position of catching the winter sun in its front rooms. I am more pleased that we have behind us open fields and copses for the children to roam about, which may improve their health.

Boulton wanted us close to him at Soho, but there was nothing suitable that was not confounded by the noise and smoke of his works. Ann will have no part of living near the 'stink' of a new manufactory, being used to her father's house among genteel folk in Glasgow. I had not noted the smoke and noise until she complained of it and now I am glad I can offer her a situation to her liking, for she has a way of winning with words well beyond me.

I have of late amused myself with a particular problem of interest to my colleagues especially Darwin who is making his own device as a copying machine. We have no good method for making copies of our many letters and drawings and much labour is spent in repetition of documents. I have used a tracing paper which, when held to the light, reveals the original exactly. I will need to work with the paper, ink and press necessary to achieve the result I foresee as quite possible and indeed very handy for us all. I have written to my invisible partner to send specimens of the most even and whitest paper and will write to Black of my success with my copying machine which I hope, will help us all with our business letters.

Perhaps it should be a comfort to me, rather than an aggravation, that Boulton is always up to some scheme or other, and assures everyone who will listen that there are great investments to be had. He lives as a man who has the Midas touch, which will soon be Necessary as our Debts are more the measure of his ambition than our earnings!

If I would let them, debts would plague me to an early death. To merely think on our lack of profit is to bring back my headaches and trembling.

*

50

'Trouble is at our door.'

Andrew Cosgrove
Handsworth, 1780

It is but a few months and Trouble is at our door. Tomas and Josiah were awaiting a visit from their family but Tomas Winterson Senior arrived unexpectedly at the house, on the New Year and in biting cold. Amy came down to tell me I had a visitor and was wanted in all possible haste. We were both at a loss to the reason for his visit since he had insisted on speaking to us both with no mention of his sons. I had first met him as a proud man in a ruffled shirt and fine coat where he readily signed the papers for Tomas and paid the Fee.

He rose to meet me pale and much reduced from his journey. He looked as if he could barely stand and seemed to have lost his gloves and hat for he had neither about him. I do not know what to do with what unfolded except to write it down so as to give it form and substance rather than melting in the kitchen air where we sat. Amy served some cider and bread, and pulled a chair up to the hearth.

Mister Winterson, what is it that has brought you to our door? Have you concerns about young Tomas and Josiah? She spoke very directly as if our good name was in question.

No at all, Mistress Cosgrove, not at all, Winterson replied. But I hardly know where to begin.

I looked at his exhausted face and I knew.

You have not come from Leicester, have you, I said.

No, he said, I have come much further. I have come from Chorley.

Amy folded her hands together as if she might lift them in prayer but she sat silent.

Have you come from the Birkacre Mill (1) where there was trouble not so long ago, I asked. I was surprised that I could feel my heart pumping so loudly in my chest.

I have, Cosgrove, and I am in a sorry way for they have taken my wife and I have not been able to free her.

I remembered the Thunder from a clear blue sky and here it was striking Winterson, pale and unable to drink his ale and his wife taken from him.

Who has taken your wife? If you do not tell us all, we cannot make sense of how to help, Amy said her voice shocked to impatience.

Winterson put his head in his hands. The poor man could not look upon another soul.

Finally, he spoke. There was more than Trouble at the Mill. There was a Riot and my dear wife Mary found our youngest daughter gone, curious to the Commotion, and followed the crowd to find her lest she be thrown down and trampled. She was swept along before she could withdraw but found our little Mary, her namesake, crying and bewildered and bought her home.

He paused and raised his head and looked at me. He looked a pale milk sop but there was a hot fury in his eyes.

She was no sooner home than one of Arkwright's Bully Boys arrived with a Constable and claimed Mary was a Rioter and he had a piece of her dress to prove she was there.

A piece of her dress, Amy raised her hands in indignation, how can a man have a piece of her dress?

Aye, said Winterson, that is what these Devils will stoop to. He had cleverly cut off a piece of Mary's dress in the crowd to have as Evidence.

We sat in silence, hardly able to comprehend such a thing.

What has come to pass, I asked, for I could not bear that Amy might start to weep.

She was taken to Chorley to await the Assizes called early to deal with as many of the rioters as they have caught. She will be charged under the Riot Act for being in a Riotous Assembly, Winterson was still as if frozen.

It seemed a long moment we sat with him deep in the Horror we had heard.

Do you need the funds to find a man of the Law, I asked finally. It broke my heart to ask a man if he could not defend his wife.

Winterson looked at me tears of shame in his eyes. If you could find a way, he said, I would be grateful to all Eternity.

My dear man, I said. I will find a way. Your sons, God Bless them, have given our Business good service. Do not trouble yourself about it.

Finally, Winterson took a draught of his ale.

You must rest, said Amy. You will stay the night with us and see the boys tomorrow.

I told Winterson I was needed at the Forge and that Mistress Cosgrove would attend to him. Winterson lifted his head and wiped his eyes. Thank you, Cosgrove, he said.

*

The next morning the boys, to their great joy, met with their father and spent some time with him, especially boasting of all they had learnt and showing him the Forge and all its tools. Winterson had recovered his strength and was a steady friend to his sons, admiring all they had done and praising their skills. I do not remember him as a warm man, but I have observed that Travails can lead to a softening of men towards their own. He indicated he wished to talk further with me and I thought it best that we took a late morning walk while the sun still had some claim upon the day. We walked for some time and then he spoke.

My family has not always been so Afflicted, he said.

Winterson, I replied, I know of what you speak. There is Change everywhere, whether we welcome it or not, and whether it will drag us all to Heaven or Hell I do not know. As I said this, I could have bitten my tongue, for Winterson at once replied.

It has driven me to Hell.

I was silent for a moment but the truth of his fortunes I desired to know.

Can you tell me what has transpired these last few years?

Winterson's words came as if every one of them was a burden.

Cosgrove, when you first took Tomas as your apprentice, we were prospering as we had done for two generations of weavers. The Wintersons had a reputation for high-quality cloth and we had much comfort in our lives, for the work pleased us and we were our own masters. We had our animals and our gardens. I am amazed to think back on it for we were in Paradise but did not know it.

That is how I perceived it, I said. I understood you were doing well in Leicester.

With the new Mills, Cosgrove, our Trade was slowly taken from us, and from all the Spinners who have for generations supplied us. You may not have heard of Arkwright's power-loom for Carding Cotton called Big Ben?

I cannot say I have, I replied.

It is the most hated machine of all machines, he stopped and grabbed my hand.

Do not think of me as a Weakling, I beg of you. I have been through hard times before when the Season has not suited us, but these are not hard times these are the End of Times. How could we compete when a new machine can take one hour to do what we can in ten?

I took his hand and pressed it. It was middle winter and a chill wind but the cold I felt was deeper than winter. I knew of what he spoke. It was the Beginning and it was the End and I had heard it said.

I thought it best that Winterson should match my pace and I walked on. We were much reduced, he continued, when Arkwright's men came calling from village to village to encourage weavers to become 'hands' in the new mills. With no work and what little silver we had well sold, our animals butchered and our stomachs moaning, I did as they advised.

There was a long silence.

I never in my life dreamed I would put my family on a dray with a horse supplied by Arkwright's men, and travel so far for so little, he said bitterly.

We have heard stories from Mrs. Cosgrove's cousins in Nottingham. The children get work but the parents do not, I said.

Yes, Winterson continued, Arkwright will have them after seven years of age and he works them twelve hours, even through the Sabbath. There is talk of a Sunday School but this is still all bluff and bluster and pity help them if they have no time keeper or do not hear the factory bell. If they are a minute over six of the clock they forfeit a day's wages! They took all four of myne and would not take me because of my age. That a man is called upon to have his children support him is beyond any Toleration. Not only is he made idle, his children grow into cherry-cheeked savages who cannot read or write.

Winterson held a kerchief to his face as if there was a sudden frozen blast upon him.

Is that the cause of the Mayhem, I asked.

How can a man stand by and let a machine take his livelihood and make slaves of his children? His voice had an anger so deep it roiled over our heads and I would have seized it myself and taken it and marched with the Rioters, on the instant.

What took place at Chorley, I asked for I did not trust myself to assume such a scenario.

Winterson spoke, his voice rough with his feeling. To protect my family, I stayed at home in our pitiful lodgings and watched from a distance as they marched by, with drums and banners flying in high good spirits for they had marched through many villages from Blackrod south and had gathered numbers over a thousand. They carried heavy sticks, which I took to be part of their Bravado.

They were armed, I asked.

Yes. Arkwright had ordered men to guard the Mill and the Military had been called but the Commander thought he had calmed the Mob and moved his men to Preston. Later Arkwright's local men fled in the face of the Mob whereupon they entered the Mill and broke as many machines as they could. They destroyed the Great Wheel and set fire to the frames and burnt the Mill to ashes.

I could hear the satisfaction in his voice and I did not blame him for it.

Now I hear Arkwright has set more than a thousand guards up around his other Mill at Cromford and has a loaded cannon set within its walls. What sort of man is he, Winterson exclaimed, that sets cannon against his workers?

A man who owns the future, I replied.

My words hung in the air and I wished I could unsay them or unthink them, but it would be as mad a Wish as turning my long clock back and saying that Tuesday was now Munday.

If it wasn't for my children I would care not tuppence for my life, Winterson said finally.

Now your girls do not have their mother with them they will need you more, I said, for God help them if they do not have a Protector.

You are right, Cosgrove, and I will not forsake them.

Winterson stopped and looked at me.

I cannot tell Tom and Josiah what has come about with their mother as yet, for they will run away from you to stand with me and I do not want that. That they are safe and have a future in your Trade is all that is holding me, Cosgrove. He looked like a man whose burden was so great it was certain to crack his very bones.

Do not fear on that account. We will take good care of them, I assured him and a strength flowed through me as if I had been given sons of my own blood.

You are both God's Good People and I am a lucky man but I have something else to ask you, he paused.

Before he could ask I assured him that whatever came to pass Mistress Cosgrove and I would treat his sons like our own. I told him that God has not seen fit to give us children and I could speak for my wife in this, for she already spoils them and delights in their presence.

Winterson shook my hand and did not speak again till we returned to the Forge where he told his sons to listen always to their Master and his Mistress, kissed them both and took his leave.

*

51

'before we could be Milked of our Ideas.'

John Smeaton
Ramsgate, 1781

The Innkeeper at Ramsgate is an excellent fellow who has done his best to give me all my creature comforts for I have endured the roads since London and beyond. I have been to Ramsgate so many times working on this damn'd Harbour myne Innkeeper has become my new brother. But the roads and tiredness aside, I have written to my dear wife that I am safe and in some relief after a visit with myne host, James Watt, at Hatchers Hill. He is a fine engineer, for that is what he has become, but he is also a secretive fellow with his eye on the main chance.

I took the young Philosopher Henry Moyes as company for I am most fond of him. He is a fast-rising protégé of Adam Smith and tells me they are both Kirkcaldy boys, which has been a great good fortune in his life. Henry has almost no Vision but, as he remarked in our chaise back to our lodgings, he sometimes wished his hearing was as poor as his sight. He is a slight young fellow but foreign to any pity upon himself. Small pox, I believe, affected his sight when just a babe, but his mind is as sharp as a new Sheffield razor!

We were bothe of the opinion that we were being fed and watered in the most congenial way before we could be Milked of our Ideas. Watt's ambition seems boundless, for he is himself well fed by Boulton's endless desire to make a name for himself on the back of whoever can lift his enterprises. He is fast becoming what Dr Johnson would call, in his genteel way, a Pompous Ass.

But I will say his engines speak for themselves. I was amazed to learn that at Chacewater, where I am familiar with the engine, his *adjustmets* as he calls them, were pumping at 341 fathoms with a stroke at 14 per minute. It is no surprise to hear that the Poldice adventurers at Truro, having witnessed with their own eyes the power of Chacewater, sing his praises.

*

As we arranged ourselves ready to partake of Watt's hospitality, I ventured to comment on the point-to-point fracas I had survived with churlish London Philosophers. My support of the theory of *Gottfried Leibniz* in his *Conservation of Energy*.(1) created an uproar. In a very short time I had several letters and several more snubs at my club by men who had suddenly decided they were Geniuses and informed me I was Inconsistent with the great Isaac Newton and his theory of the *Conservation of Momentum*. I am a man of practical application and my experience led me unequivocally to the conclusion that Leibniz's work was the most useful. The protests of such addlepates heightened my certainty.

I thought I would speak to an ally, but no, Mister Watt, Engineer, was not inclined to follow the same principles. I was amazed to hear him pronounce that the Germans were often dogmatic and that the English mind was exceedingly flexible and between Leibniz and Newton he would take Newton.

Moyes nudged me to caution as he feigned a thoughtful consideration of Watt's sudden outburst of Cant.

But I could not let it rest. With the greatest respect to Newton, I responded, I think even he would acknowledge that Mathematics likes to have a practical application. It likes to travel and has an uncanny ability to arrive the same as when it left. Of course, Scottish Mathematics may be the exception, I added.

I watched Watt flush with annoyance and, being duty bound as Host, hold himself in check. Then he laughed, because he is, at heart, a most genial fellow.

I bow to your greater knowledge in these matters, Smeaton, he said. Moyes audibly sighed in relief.

Of course, he said, offering to be peacemaker. Both learn'd men could be right.

Watt, I must admit, paused as did I. Before either of us could respond Moyes continued.

Mr Watt, Sir, I am most curious to learn about your latest experiments with your steam engine. He asked in the way a butcher might enquire as to the health of a farmer's pigs.

Watt turned to us and looked quite dismal.

I have a double acting piston and an engine that can push as well as pull, he told us.

I would have been a proud man to have made such an announcement but Watt seemed troubled.

You will head for rotary motion then? It is rotary motion that is all the go. I am myself approached every month by some miller or manufacturer asking if I could supply him with a rotary machine, I asked.

It is a devil of a thing, said Watt. Boulton tells me he is also besieged.

He says the people of London, Manchester and Birmingham are steam mill mad.

But surely, I have great sympathy for these cities, Moyes said. They cannot shift their mills from the rivers. They cannot tell their horses not to eat.

What else can you do, I continued. There is no other way to progress but if you can make a jerky Fire-Engine produce a circular motion as reliable as a turning water wheel I will Salute you.

Watt laughed at the thought of me offering a friendly kiss.

Then you may have to Salute me, Smeaton, in the foreign way, for that is my intention.

Moyes lent forward as if he could read every expression on Watt's face.

A complication with your progress has recently arisen, he ventured.

If you are referring to Wasbrough and Pickard we have already been in negotiations with them. I have seen their Snow Hill engine and it was a caw-handed thing until Pickard fitted a crank and flywheel, Watt said at once.

And they had the acumen to get a Patent post haste, said Moyes, for Wasbrough's engine already turns a wheel.

Indeed, said Watt, I have been well bolted out.

That is a fix you are in then, for it would not be politic to challenge the Patent, I said.

The Foxes would be out for me if I did, Watt declared, even though we have a weighty suspicion the design was stolen from us.

How so, said Moyes, eager to learn of any such gossip as I have found common in philosophers.

We have put it down to our worker Dick Cartwright at the 'Waggon and Horses' making merry with one of my designs by way of chalk on a table top. He is not usually a toss pot but he was well jagged that night.

Do not be too hard on the man Watt, I said, he was probably boasting of the prowess of Boulton & Watt.

If he had kept his mouth shut we would have even more prowess, Watt scoffed, besides, he continued, the crank is a common day device used in every spinning wheel in the country. It can be seen on every grind-stone turned by hand and every knife-grinder's foot lathe. Watt's voice rose with his frustration.

There are many variables for you to consider, I ventured.

Everything is variable, Watt replied. As Pickard has found applying a crank to a steam engine is taking a cheese knife to cut bread.

If you cannot use the crank you are left with some geering device, I ventured.

That is the way of it, Watt said.

And that may be the way, I continued, to resolve the conflicting actions of the beam and the piston.

I would say the crank design is the superior to all other devices in both savings and movement, I continued.

Watt bristled as I expected he might.

That may be so but Wasbrough and Pickard will not have the field all to themselves. They seek to have a marriage between our engines and offer me their crank in return but I will have none of it. We will match them and then better them. I have in mind a solution which may serve to outwit these Snow Hill engineers.

He waited for my response.

It will be a geering solution, I repeated and Watt nodded.

We fell silent, each musing on their own thoughts.

I knew what Watt faced. It is never a small thing to progress with new ideas; they come in a flash, appear simple, delight the mind, seduce your funds, become a horror of Complexity, and if resolved have the Temerity to Appear as if they were always so. I was Intrigued but I did not envy Watt the work ahead.

Another silence ensued and I wondered if Watt was waiting for me to empty my mind into it. Instead, Moyes emptied the last of his cider tankard and tapped his fingers on his chair.

Watt, sir, I am amazed by all I have heard, but am tiring as is usual for me. I must prevail on John to see me home before I embarrass myself and fall quite asleep.

Of course, said Watt, springing to his feet. How thoughtless of me. When engine men get together all civilisation is left behind.

*

Moyes was more lively on his way back to his lodgings.

I wonder, he said, if Boulton & Watt will survive the end of Boulton & Fothergill?

I knew of what he spoke, as it was no secret that Fothergill was tired of Boulton's extravagance and the long partnership was in most acrimonious communication. It seemed most unlikely that Boulton would be so unwise as to also dispense with James Watt. I was about to say so when Moyes continued.

I have a fair sense of Watt's plans, he mused.

How so, I enquired,

There is talk that William Murdoch, the Soho engineer, has devised a geering arrangement to serve in place of a crank.

Murdoch has a fine grasp of engineering. Watt is lucky to have him, I replied. Moyes paused, for he knew I did not have much truck with gossip. You know, John, he continued, I did mean I was amazed at your conversation.

That Watt was trying to pick my brains instead of my pocket, I asked.

No, not at all. You are both so well versed in these extraordinary ideas I think you have forgotten how strange they may sound.

How so?

You engineers are brave fellows to challenge wind and water. You are building machines you say will beat them both. You plan to give the world a wheel that turns without a man or beast.

The world is already full of wheels, I replied being tired and preferring quiet.

But Moyes persisted. But this is a turning wheel which can be set anywhere its owner desires and will never tire as long as we have coal and water aplenty.

He gave a short laugh of excitement and I did feel energy stir within me. Watt may be a prickly fusspot but he is running his own race and while many can match his pace no one will match his results, with or without Murdoch. I knew he would be at the Patent Courts within months and some new design would advance Boulton & Watt.

It is momentous, Moyes continued on his theme, but I wonder at its consequences? You are the philosopher Henry, I replied, and I a mere engineer, so I will leave that for you to ponder. Moyes mused upon this and we drove home in peace.

*

London, June, 1781

I have this day received a request from the Commissioners of the Navy, an honour I cannot refuse. Since Boulton & Watt decline to partner with Wasbrough & Pickard the Commissioners are seeking what they call an 'expert' opinion on which of the two engines may be the better for the grinding of flour at the Deptford Victualling Yard. This is an onerous task since Wasbrough has been at work on his engine since early January and had been assured of his acceptance. The Navy Commissioners are having second thoughts and have made it a Contest far too late for my comfort. In fairness to bothe parties, and to my general belief that steam is not as reliable as the power of water for milling, I may well find it politic to favour neither engine.

*

Felix Farleys Bristol Journal

Sunday Oct. 21st 1781, died, deeply regretted by his friends and acquaintance, Mr. Matthew Wasbrough. The public have to deplore in him the loss of one of the finest mechanics in the kingdom, whose early genius brought to perfection that long-wished for desideratum, the applying the steam engine to rotular movements. Upon these principles, he lived long enough to complete several ingenious and complete pieces of mechanism, of which the corn and flour mills of Messrs. Young & Co., in Lewens Mead are striking monuments of his extensive abilities. His name, therefore, will be handed down with veneration to the latest posterity.

Original Document: *Felix Farleys Bristol Journal* 1781 Obituary for Matthew Wasbrough.

1782 Specification of Patent

SPECIFICATTON OF PATENT, MARCH 12TH, 1782, FOR CERTATN NEW IMPROVEMENTS UPON STEAM OR FIRE ENGINES FOR RAISING WATER, AND OTHER MECHAN'CAL PURPOSES, AND CERTAIN NEW PIECES OR; MECHANISM APPLICABLE TO THE SAME.

To ALL TO WHOM these presents shall come, I, JAMES WATT, of Birmingham, in the county of Warwick, Engineer, send greeting.

WHEREAS His Most Excellent Majesty King George the Third, by His Letters Patent, under the Great Seal of Great Britain, bearing date at Westminster, the twelfth day of March, in the twenty second year of His reign, did give and grant unto me, the said JAMES WATT, His especial licence, full power, sole privilege and authority, that I, the said JAMES WATT, my executors, administrators and assigns, should, and lawfully might, during the term of years therein expressed, make, use, exercise, and vend, within that part of his Majesty's Kingdom of Great Britain called England, his Dominion of Wales, and Town of Berwick upon Tweed, my invention of 'CERTATN NEW IMPROVEMENTS UPON STEAM OR FTRE ENGTNES FOR RAISING WATER, AND OTHER MECHANICAL PURPOSES, AND CERTAIN NEW PIECES OF MECHANISM APPLICABLE TO THE SAME;' in which said recited Letters Patent is contained a Proviso obliging me, the said JAMES WATT, by an instrument in writing under my hand and seal, to cause a particular description of the nature of my said invention, and in what manner the same is to be performed, to be inrolled in his Majesty's High Court of Chancery within four calendar months after the date of the said Letters Patent, as in and by the said Letters Patent, relation being "hereunto had, may more at large appear.

Original Document: James Watt 1782 Specification of Patent

52

'We sat amazed as if we had been turned to stone'

Andrew Cosgrove
Handsworth, 1783

We have kept Tom and Josiah as best we are able in these last years of cursed wars and ebb tide bankers and coins thin in the pockets of our customers. Winterson has done his duty and written to them from time to time with news of the family, although he has not mentioned their mother's fate which was as terrible as it was unjust.

Mary Winterson was given no leniency at trial as a mother or as an innocent bystander caught up in the Tumult. The lawyer we had paid for told Winterson there was little he could do as the Riot Act had grown in harshness over recent years and presently spoke of Severe and Effectual punishment, and that being a mother and so responsible to uphold her lawful duties to her children, the Judges were set to be harder upon her than other prisoners before them.

She was given two years' hard labour and because there was no women's prison under Lancashire jurisdiction and she could not be sent to the American Colonies, she was sent down to the Thames at Woolwich. The poor woman served her time on the navy ship *Censor*, made a hulk floating in tidal rubbish. She cooked and cleaned for the guards and survived only because of the Kindness of the Lawyer who was so affected by her Innocence he arranged his fee be used for the comfort of those Guards to protect Mary as best they could.

She returned home, according to Winterson, much changed and deep in Melancholy and, despite the efforts of her girls, she passed to God several weeks later. Since then his youngest girl, aged eleven, has also passed to the Lord from a lung disease brought about by working in humid conditions made to suit the cotton but not the children leaving straight into the cold of evening.

I do not know how it can be that Winterson is so short of shillings he cannot supply blankets and shawls to protect his girls! He has written this last week to say that he will be visiting us as it is seven years since Tom started his apprenticeship and his Papers should now be signed. They have been waiting, signed and stamped, these last months for Winterson to come and collect his son.

That son, Tomas Winterson, is well certified to join London's Worshipful Company of Blacksmiths if he has a mind. He has been out shoeing and gate-mending as his own Master for many weeks. I clearly stated that he is proficient in all manner of tools necessary to satisfy the needs of farmers, hunters and housekeepers, and professed that his manners and morals are of the highest quality and no man has reported to me otherwise.

*

Having been out attending to gates on a farm, I arrived home this early afternoon for lunch with Amy to find her most ruffled as Winterson had arrived and was at the Forge seeing his boys. Her eyes were wide with concern when she told me that he was not alone and had what she glimpsed as a Quaker Priest with him.

I sat down for my lunch and joked to Amy that Quakers did not have priests and if it was a Quaker he was more likely to be an Entrepreneur looking to buy us out. Amy did not think me droll and turned to me in some anguish. They have come to take the boys, she said.

They have come to take Tom, for his apprenticeship is well done and I signed his Papers months ago, I replied keeping my voice steady, for now the time was upon us I too felt a tearing in my chest.

Will they take Josiah also, she asked her hands busy moving pots from one place to another on her hearth.

I do not know, but he is only a few years short of his Papers and I think his father is still keen that his boys have a Trade, I spoke more in hope than certainty.

She stood and looked at me, grey and stooped as if time had taken away ten years of her life.

I do not think I could bear to lose them bothe, she said.

Neither could I, dear wife, I said. I will go to the Forge and see what Winterson and his companion are about, which I did.

*

It was some time later we came back to the house. The five of us walked through the cold air of late afternoon and the Quaker remarked that soon old man winter would be upon us now it was November. The boy's faces were flushed from the warmth of the Forge, but they walked like they were in a dream and I did not think it likely they noticed much of anything.

We entered the warmth of the house and I was never more proud of Amy in her fresh apron, a respectable table offered and a good fire going. I introduced her to Quaker Smith who bowed in great respect and Winterson took her hand in the warmest way.

She sat the boys down and served them soup and offered the men an Ale, to which they bothe shook their heads and wondered if she could provide them with tea. I poured myself an ale, sat down beside my own hearth and said to Winterson he should inform Mistress Cosgrove of his news.

Mistress Cosgrove, he began, I cannot begin to thank you for your care of my boys. They are in robust health and sharp in their wits. I will send prayers to bless you all the days of my life. He stopped, choked with unshed tears, his voice a soft rasping sound as if his words were a saw cutting across his very soul.

Dear sir, she replied, it has been our great pleasure.

Winterson remained silent and stared ahead of him as a man would who has just woken from a deep sleep.

Mistress Cosgrove would like to hear the rest of your news, I prompted him.

Yes, yes, he said, lifting his head like an animal might at the sound of a whip.

I have some good news at last, he said, for I have joined the Society of Friends and have become a Quaker and Elder Smith has become my friend and mentor. The Winterson family is most blessed for we have been offered passage to Pennsylvania where we will find a much better life than the one we have at any manufactory in Cromford. My grandfather came from the land and we will go back to it and provide for our communities in Philadelphia. Pennsylvania? Amy said it with such disbelief Winterson could have been talking about the planet Mars.

Yes, indeed, Mistress Cosgrove, the proud state of an Independent land where the Liberty Bell has rung out its message of freedom, Elder Smith interrupted with a grand gesture.

There was another silence as we took in Elder Smith's words. He did not speak again as the silence grew and he reconsidered what might be heard as a treasonable delight in his country's recent defeat. The boys had their heads down over their soup like cowered dogs.

I could not bear to see them so crushed. They had been told news of their mother's and sister's death and the prospect of life in a land foreign to them. I could have cursed Winterson for his absence, his silences and now his fulsome melodrama.

Mrs. Cosgrove would like to know what plans there are for Josiah, I said.

Ah yes, said Elder Smith, Josiah will stay to finish his apprenticeship and then, through our Fellowship, he can join his family in Philadelphia.

That is a most reasonable plan, I ventured, for I thought at any moment Winterson was about to say something different.

He leant forward his hands clasped tightly together as if he was under some restraint invisible to all but himself. It was then that Tom stood from the table and came towards his father. He did not once look at Elder Smith. I noticed, to my shame, how poorly his clothes looked, and how his growing toes might burst from his old shoes in front of our eyes.

Father, he said, in a clear voice as if he had been rehearsing his speech for many months, I will not come to Philadelphia. I do not plan to be a blacksmith. I plan to be an officer in His Majesty's Navy.

He stood as tall as he possibly could and waited for his father's reply.

Winterson looked at him dumbfounded as we all did.

I do not think that will be possible, said Elder Smith.

Smithy Cosgrove, sir, would you not say that it is Josiah who has the true talent for smiting and creating iron work? Tom looked at me in such a confident forthright fashion I found myself answering without regard to his father.

Tom, I said, I would be hard pressed to say which of you excels. But lad, you cannot be thinking of the Navy. It would be too difficult. Where did you get such an idea?

I have these last seven years worked were the village comes for warmth, for problems solved. I have heard many tales exchanged, and much gossip while they wait with for horses, or their pots and pans. Talk is, we will be having another war with the French, as we always have, and I want to join and fight.

Josiah then came and stood by his brother.

Dear Master, he said, addressing myself, Tom says you are a friend to James Watt from the Soho Manufactory and should be able, when I finish my apprenticeship, introduce me to a man who can further my career. His voice had a sing-song lilt as if he had been practicing a poem to impress a school master.

We sat amazed as if we had been turned to stone and had lost the power of speech.

Amy was the first to break the silence.

How much money have you saved, Tom, she asked, for we have given you clothes money.

Tom smiled for the first time. I have saved every farthing I have ever seen and have a purse of twelve Pounds. I have letters from my customers to date saying they would be proud to have me serve in His Majesty's Navy.

I was flabbergasted. You have such letters?

Yes sir, I have. Tom stood very still, waiting.

Winterson spoke, his voice barely as whisper. And how would go about convincing the Navy of your worth? You have never seen the sea.

I will apply to be a servant for an officer. After three years I will be a midshipman. I have been practising my reading and writing and Mrs. Cosgrove declares my Mathematics well beyond hers. Tom glanced a shy smile at Amy.

My wife raised her hands appearing helpless before his certainty.

That is true, she said, Tom has a natural aptitude for Mathematics. He has moved well beyond my teaching.

My mathematics are well beyond anything imaginable, said Josiah and suddenly, as if a cork out of a bottle, we all laughed.

I stood as Master of the house.

There is a lot to consider gentlemen. After you have eaten you will want to return to your lodgings in Handsworth. I suggest we sleep on what we have heard and speak again in the morning.

Winterson looked relieved. Thank you, Cosgrove. That will be most suitable.

*

53

'the most advanced and efficient engine ever built'

James Watt
Fellow of the Royal Society of London
Fellow of the Royal Society of Edinburgh
London, 1784

Boulton and I conversed with Samuel Whitbread in a room he called his office, a large sunny room away from the noise of the brewing. We were seated most comfortably and the Brewer served a fragrant tea and I thought it a most felicitous equation that we drank from Wedgewood's cups. To our satisfaction Whitbread is a friend of progress and experimentation, but he is not a man for equivocation.

I hope you enjoy my chatter broth, be began, for it has arrived this very day, fresh from Billingsgate, although China is a bit further! He laughed, but did not stop for an answer. He looked directly at Boulton as if he might seize upon his person and shake him.

I don't know how you Birmingham men have done it, he said. They say Watt is a genius but you and I Boulton scrap for every penny. So I ask you directly, can your engine do all you say it can? It will be my second engine, and you say it will be better than the first?

Boulton was not his usual self but expanded in his usual way. Our answer is yes and more, he said as if it were set in the tablets of Moses!

The engine we supply will be the most advanced and efficient engine ever built. It is no simple pump, it is a rotary engine which will have 10 revolutions per minute, and every minute of the day.

Every minute of the day, Whitbread queried, as if he might laugh in our faces.

Not only will it be the most efficient engine ever made, it will save you many hundreds of pounds in the annum, I interjected. I wondered if Whitbread knew Boulton had suffered, and most suddenly, the loss his dear wife in the terrible heat of last summer.

If Watt says so, I can assure you, it is as good as done, Boulton said, much pleased with my interjection.

It will have our revolutionary sun and planet gears to drive the flywheel. It will serve your mill day and night. It will not sleep or come late for work, I declared.

Whitbread smiled at last, then it will fit in well, for I do not have workers who come late to work. They do not want to pay my fines. He offered us more tea and happily signed our contract.

*

London, 1785

I have installed a rotary engine at Chiswell Street Brewery for Samuel Whitbread to replace his old horse wheel. Whitbread loved his Clydesdales and would not dispense with them completely. He wanted the horse wheel retained, more I think out of sentiment than doubt in our engine. We now measure Savings, not in coal, but in horses. Our engines replace teams of horses and so we needed a calculation to measure the working power of a horse. Boulton and I have now settled on a measurement we are contented with. He insisted we measure a brewery horse using the same equation I have used many times. We measured the amount in pounds weight a horse can pull over one foot of distance in one minute. As I have previously calculated, it was 32,400 foot-pounds, which we called for the ease of it, 33,000 foot-pounds.

We accommodated Whitbread's engine in his old stable. Boulton heard it excited his interest so much, Whitbread would not have it started without his presence! It has a 24 inch cylinder with a 6 foot stroke and a riveted copper haystack boiler with steam pressure of 5 psi. The piston drives a laminated wooden beam which I have adopted for better strength and cost than a trussed beam. Adjoining the stable was the malt mill, where a 27 foot diameter horizontal wheel attached to a shaft drove the malt millstones. I connected the engine by a series of wooden line shafts to drive the mill shaft and replace four horses. Whitbread may love his horses but no horse can prevail over steam.

By way of a reciprocating pump connected to the engine's beam, I was able to lift water from a local well to a tank on the roof of the brewery. The engine turned an ingenious Archimedes screw to lift the crushed malt into a hopper. (I would not be the first to call Archimedes ingenious, nor use his design) Whitbread needed a hoist for lifting other items as necessary and that was easily done. The engine has the power to drive a three-piston pump for pumping beer, and a stirrer within a vat. By adjusting the diameter of the gears, I kept the speed of the wheel consistent.

Whitbread insisted the engine do its best without further disturbance in the work rhythm of his men. I do not like frightened horses, he said, and I like frightened men even less!

*

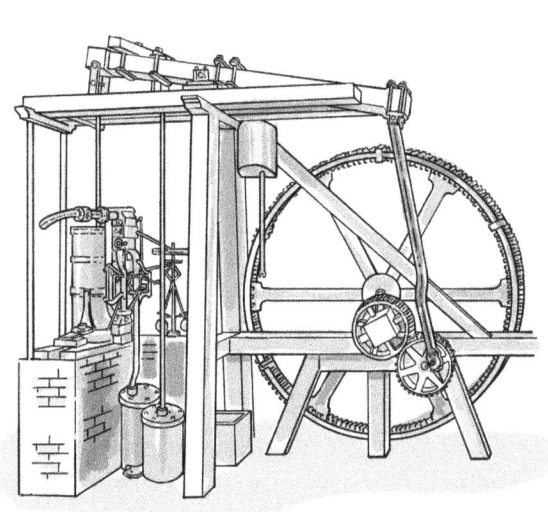

James Watt Steam Engine with sun and planet gears 1785

54

'the Devil himself was loose upon the road'

James Watt
Birmingham, 1786

William Murdoch has sent me news from Cornwall and he has set my mind at sixes and sevens. He writes of the adventures of his three-wheeler coach-car in its testing, which he conducted, most advisedly, late at night on an open road near Redruth.

It is easy to envisage the racket and storm cloud it would make and how, to come upon such a contraption under God's starry sky, might concern even the bravest soul. He writes that it was his luck to encounter a clergyman enjoying an evening walk upon the same road and that now Redruth has heard that the Devil himself was loose upon the road and no man should walk alone at night. Murdoch has a ready pen and he made me laugh at his story but I do hope that the Commotion he caused may give him pause as to future experiments.

It is not the first time I have contemplated the notion of steam-on-wheels. John Robison and dear Darwin, plus Edgewoth and Boulton himself, endlessly dream and plan on scraps of paper left for me to puzzle over.

That steam could power some kind of vehicle, especially if it is set on rails for smooth running, is the most Vexed idea ever to be discussed in civil engineering. I am aware of its novelty and the excitement of standing on such a moving vehicle. The possibility of moving considerable loads across suitable terrain would delight many Adventurers but they would not be so happy at the dangers such steam power possesses. I have become more and more wary of the power of strong steam and the problem of containing its power with a trustworthy boiler.

It may disappoint Murdoch who amuses himself with such distractions in the wilds of Cornwall, but I will write that I do not favour such contraptions because of the dangers of explosion which may injure many. Boulton says I am too cautious, but the pressure of steam is not a toy to be played with and I will not have the death of any man on my conscience. I will beg him to heed my advice and take up some other adventure and experiment to ease his boredom and entertain his ingenious mind.

Murdoch might not know I described such an engine in the fourth article of the specification of my patent of '69; and again in the specification of another patent in '84, together with a mode of applying it to the moving of wheel-carriages. He will no doubt desist as he is aware that my Patent over use of such steam holds until 1800.

55

'at a loss as to their status, being neither
plants, animals or fossils'

Erasmus Darwin
Derby, 1794

I have much to reflect upon and all with great satisfaction. After the publication of my poem *The Loves of the Plants*,(1) which was well-received, I was strangely not released to Liberation but whatever weather or trouble was sent to Derbyshire, and wherever it found me, I scribbled notes. If one thing be true of my dear Lunaticks, now much dispersed, we are great scribblers. There would hardly be a day when we have not written one to the other or the other to yet another!

I have amassed a veritable pyramid list of wonders among which I count my dear wife Elizabeth, my marvellous botanical muse and mother of our children. She has also become a great guardian of my health and since she has sworn me off the Devil's brew she says I have gained twenty years of virility and smiles happily to herself. She has been my delightful companion through my *Loves of the Plants* and has urged me to follow the Swedish botanist, Carl Linnaeus,(2) who has gifted the world a wonderfully clear system of classification for all organisms.

As all men are susceptible to flattery I confess the warm applause I received added to my satisfaction and my bank funds in equal proportion. To have contributed to Science is an honour but to have contributed to the achievements of what I used to call my Birmingham Philosophers, my circle of dear Lunarticks, is a great joy. This month, at last, *Zoonomia*, (3) is published, and like a ship in the eye of some carefree storm, I am momentarily at rest. My laws of Organic Life are presently circulating in the Digestive Systems of philosophers and critics alike and

I await news of their Colic or other Maladys blamed on my theories. I await especially, a response from those few not given to the fictions of fancy, in particular, those who see that Ignorance and Credulity have always marched together and that we philosophers stand on islands in the seas of Perpetual Error. Linnaeus may have classified our species as *homo sapiens*, being wise man, but I fear it may have been more in hope than realisation.

Boulton and Watt, who have become enlarged versions of themselves now they have achieved recognition and indeed, fame, still give me joy. If Boulton was a man of confident assertions he is now a grand master of it. He wrote a description of his engines to include in my third edition of *The Loves of the Plants*, under the modest title, 'Catalogue of Facts'[4].

Convenience and properties of B&W improved Engines.
- a. *it is the most powerful Machine in the World*
- b. *it is the most tractable*
- c. *it is the most regular and hath the property in itself of conforming to the least work.*[4]

This brought to mind a memory from years ago when Boulton and Wedgwood were firing merry guns directing the campaign to win support for Duke of Bridgewater's canals. Wedgwood persuaded me to be the author of his pamphlet commending the Canals investment potential, but was short on supplying me with information and famously remarked that if I had run short on facts I must create some! Nothing stands in the way of such men as Boulton and Wedgwood, and I have been happy to trim my feathers as their whirlwinds pass by. I hear now our dear potter is withdrawn to Etruria Hall, where he struggles with his Ailments, and I have already written to him to encourage him to find a proper Physician for his Suffering rather than use the Apothecaries he provided for his workers. He called himself a simple potter though history should know him as much more, for he, before all men, joined creativity and moral judgment together in a manifestation I wear proudly on my lapel. It is his Medallion of '88 where a noble slave in chains appeals for his freedom. 'Am I Not A Man And A Brother?' he asks, begging other men that he be free of his slavery. I sadden at the thought that Wedgewood, who has done so much to answer such a plea, may not live to see its triumph.

It is said of literary men they scorn to know the second law of thermodynamics but it cannot be said of the Lunar Society, who thrive on such knowledge. We are not closed to philosophies of our time and are of such liberal tastes we lead society forward with audacity and fortitude! I say this as a simple truth rather than a righteous boast for Boulton, Watt, Priestly, Kier, Galton and myself have never shyed from supporting Wedgewood, Clarkson and his friends in such a cause. (although, if I am an honest Journalist, it is also true to say that as men of business Boulton & Watt have not demurred from selling their engines to any plantation owner who has the liking for steam power but who has not? It is the terrible truth of our times that Commerce and Slavery are bound one upon the other.)

Watt, always a man at the mercy of his anxiety, now spends his time fuming over Patent Battles he has lost, or won, since neither state assures Royalties flow the way they should. He writes constantly to me of well-dressed vagabonds who have come to plunder his designs. He adds grimly that it is always inconvenient for a thief when his victim keeps his breeches pocket tight buttoned. Moreover, he studies Patent Law as if he is about to don a wig and head for the Inns of Court! He tells me that thieves are all about and though they may not consider themselves *mischievous to the State* as patent law prohibits, they are mischievous to him. He will let no man feast from his labours. He is no Arkwright who will grind men into the dust but he worries me as he is worn thin by such crusades. He is after the Cornish mine captains who, it must be said, do moan and spit about their dues as if they had never signed a contract or never had advantage of another's enterprise. Presently Boulton & Watt are much put upon by Edward Bull, their former worker, who joined with Richard Trevithick Jnr to design 'inverted' engines, their design by another name, and now fights them through the courts. They are this week in the Court of Common Pleas before Lord Chief Justice Eyre and I hear Dr Black, Robison and Roebuck have promised their support.

It cheered me greatly to see Watt's whimsy still bubbles along keeping time at bay, for he did write a grand account of his research on the matter of classification. Regarding the placing of steam-engines among the plants he quotes nothing less than Linneaus' *Systema Naturae*. He worries that he cannot find steam-engines in Linneaus' systems and so is at a loss as to their status, being neither plants, animals nor fossils. He writes a jesting truth of the matter, for this century has seen the growth of engines of strength and endurance far beyond anything imagined in our boyhoods, and no Sage knows yet what to make of them. For Boulton they are units of pleasure as measured in profit and acclaim; for Watt they are less and less troublesome urchins and more a certain platform to his status as a leading engineer. He would never say such a thing, but since Smeaton's death, which saddened the whole country, Watt is the leading engineer in the Kingdom.

*

The day has been warm and the wind a sweet zephyr perfumed with jasmine to lull me to nap away the afternoon. If I could call the King of the Elves to sing for me and sit with his harp upon my pile of books, I would have him begin with *E conchis omnia*, [5] as I wrote ages past, for it is true, colic in my readers or no, that life has risen from shells, meaning the sea is the soup of all life. If my life's work is summed as one thesis it would be that we are all from one living filament, possessing the faculty of improvement by our own inherent activity and of delivering down those improvements by generation to our posterity, world without end.

*

56

'I would cheer you too except you do not come cheap'

James Watt
Member Smeatonian Society of Civil Engineers
London, 1795

Today was a day I will long recall in the making of engines.

I returned to Whitbread's Chiswell Street Brewery by way of the Thames where the crowds made me and the lads stop and have a keek for we had never seen anything like it. The river had frozen solid, as has most of the Kingdom, and London was making merry on it. Ice had dammed the old bridge and created a solid ground for all sorts. Shops and skating rinks, bowling matches, and all kinds of whatnots flaunted their wares curtesy of Father Frost. It was all I could do to persuade the lads to keep with me to Chiswell Street, which I did on the promise of many a fine ale from the man himself.

We were to begin the measurements for the brewery's new engine, thinking as usual no worker would lift his head to notice a fussy engineer come to tinker with what should be left alone. The noise and general hum of activity was at first an assault on our ears, for we were amazed at the expansion of the works, but then another noise prevailed. As we approached the men about stopped work and raised a cheer to Mister Watt, Engineer, and doffed their caps. Their voices echoed through the milling halls and seemed to bounce at me from Whitbread's copper vats and slide off the sides of the hops pockets and run about my legs. I stood as a man struck dumb, and could hardly raise my hat in acknowledgment, which I eventually did, to further cheering. As Whitbread himself told me later as we sipped his favourite chatter broth to warm ourselves, my engine had expanded his business, he had employed more and lightened the heavy workloads of the past.

Why should they not cheer, he said raising his voice, I would cheer you too except you do not come cheap.

Whitbread, do not joke me, I replied, you are raking in your Fortune and everywhere I go in London I see Whitbread Ales.

That may be the truth of it, he replied. But you pull up a fraction short Watt. We are fast becoming the largest brewer in the Nation. Of course, the Gin-shops hate us for we have our own hop gardens in Kent and offer the best Porter in the Kingdom. And you, dear sir, are hardly in poor street. I hear you have forty engines alone in Cornwall and dozens more all over the counties multiplying by the week!

He laughed at my expression, and patted his ample stomach most satisfied but then lent forward abruptly and smacked his palm on his table.

What can you do to give me more power? I do not want you mucking my engine about, but I want more power than the old grinding millstones.

That is exactly the reason for my visit, I replied. I am in preparation to make your engine *a la mode*[1] in engineering and double your power.

A La Mode? Whitbread snorted. You sound like Boulton with a few Ports in his belly.

Well, as you know Whitbread, I replied, Boulton's manner is infectious. He can turn a mole-hill into a mountain and be the one to climb it.

But you are not offering me an expensive mole-hill, quizzed Whitbread.

Much, much more, I quickly replied. We are offering a new 25 in. well-bored cylinder, new steam chests and valves. The engine will have 20 revolutions per minute and create a steam pressure of 10 lbs. per square inch with the same 6 feet stroke. It will be the best made engine of its type in the world with power of sixty horses.

Whitbread sipped his tea.

Boulton has long had a similar rotary engine which he calls his 'Lap' engine, I continued. It has successfully driven forty-three metal polishing machines without hesitation for many years but your engine will include our latest improvements.

And what are the improvements?

You will have a double acting engine with a piston that gives power in both directions of its movement, that is, it pushes the beam up as well as pulling it down. Our parallel motion device keeps the piston rod vertical and thus we are able to seal the cylinder at the top. We can double the power without increasing the cylinder size. Your engine, Whitbread, is driven by the most astonishing power of simple steam!

A steam engine is what I consider I am buying, said Whitbread dryly. Since Boulton had often called him London's richest Curmudgeon I smiled in agreement but I would not let him gainsay the engine.

Your new engine will have all that Boulton & Watt can exclusively offer, I replied, for Whitbread, sir, you must agree there are no others who can match us.

Whitbread laughed. Wasn't that the engine at Boulton's Albion Mill in Southwark that burnt to the ground and made your competitors happy for miles around? What did they sing - Success to the Mills of Albion, but no Albion Mill?

I nodded in agreement, for it was true and I could not deny it.

That was the one, I replied, and continued, Samuel Whitbread strives to give the working man beer. Boulton & Watt wanted to give him bread. If our competitors sang with delight it was because we outshone every flour mill in London.

Hmmm, said Whitbread. They were picking up the burnt remains of your mill in St James' Park I believe.

It was a disaster, I admitted, but the fault entirely of a negligent fireman and not the engine. It was an engine, Whitbread, which had fanned, sifted and dressed the meal and lowered it onto river barges in record time. It drove 20 pair of millstones and ground 9 bushels of corn in an hour. Sales were at £6,000 per week.

And will you give me such a profit increase, he replied fixing me a look like a Bailiff come to collect my arrears.

Indeed Whitbread, this new double action rotary engine will assist every process in right good time.

So it will add to my production and decrease my costs without blowing the roof off into Chiswell Street, Whitbread inquired with a doleful grimace.

That will certainly be the case, I replied, not rising to his bait.

There will be no other engine in London, or the whole of our Isles which can boast of all of our designs. Besides our condenser and Sun & Planet gear, it has of course, parallel motion, which I have explained.

Yes, I have heard it spoken about around the clubs.

I was most heartened to hear this for my parallel motion gives me much satisfaction and I am often drawn to watch its motion.

Then you may have heard I have added an additional pantograph linkage to the design, I continued. This allows an engine to be smaller because the linkage will be more compact.

You will charge me more but give me a smaller engine, Whitbread countered and spluttered his Tea.

That is the way of progress, Whitbread. But by way of compensation you will have much more power, nay, it will double your power. You will hear the difference for the engine will have a more regular motion, I explained.

You ran into trouble with the crank some years ago, Whitbread enquired.

The Sun and Planet gears are independent of the drive torque, I said. There is no crank involved.

Whitbread nodded over the top of what I recall Wedgewood describing as his *Butterfly Bloom* teacup. Whitbread caught my eye and explained his cup.

My daughter Mary's choice, he said, most proudly. She is trying to make me genteel, he explained.

Well Whitbread, sir, the cup is certainly genteel.

He came close to a happy smile. A pity you never could use the crank, he said.

Boulton & Watt are well ahead of the simple crank, I assured him. We have made considerable advances. You may have heard, I added, for I was liking my recitation, we use the expansion principle, that is, a valve to cut off steam before completion of the piston stroke. This had the effect of saving fuel which is to your great benefit. We have also an engineering Curiosity. When both Sun and Planet have the same number of teeth, the drive shaft completes two revolutions for each double stroke of the beam instead of one. Extra value, I added, for I had become as handy as Boulton in praising Soho engines.

That would be much to my liking, he replied. But have you considered fitting some stronger teeth to your gears than wood?

I should have been prepared for Whitbread's acumen for he was an old mill man and he had me going through my paces. I assured him I had devised brass covering for the gear teeth.

You may see another device new to your eyes, I continued.

Indeed, said Whitbread, more *a la mode?*

You will see indicators of the steam pressure inside the cylinder. I will attach barometers.

To measure the pressure of the steam inside the cylinder, he asked and raised his eyebrows most quizzical.

Yes, these pressure gauges will look like a little piston with a spring attachment.

Good God Watt, that is most particular. He slapped his thigh most impressed.

We will also fit another innovation, I couldn't help myself for I had a mad desire to confound him further.

And what does this new device do apart from add to the price, Whitbread asked.

I have called it a Centrifugal Governor. Any of your men who have worked in cotton mills may recognise it.

Aye, said Whitbread, it is a regulator. I recall we used a mechanism to regulate millstones. An odd codger. I thought it was a Dutch device?

He was as sharp as a tack as my men would have said.

It was. Christiaan Huygens, the Dutch philosopher long since departed, invented the device, I admitted.

But it will take a Scot to make it work on my machine, Whitbread replied in better humour.

It is an adaption, I explained, a conical pendulum governor. It will ensure the engine runs at a constant speed, which is good for the health of the engine and for the health of your account books, as this will also save your fuel. It has been tested in our engine at the Soho works and is most effective.

Whitbread lent forward a question on his lips. It will be connected to a throttle valve, will it not, that regulates the steam powering the engine?

Indeed, it will.

And it will rotate faster as the engine speed increases and those lever arms will pull down finally on your throttle valve?

Yes, it will, Whitbread. We call it a thrust bearing, which moves a beam linkage, which reduces the aperture or opening of a throttle valve. The rate of steam entering the cylinder is thus reduced and the speed of the engine is controlled, preventing over-speeding. So yes, you are correct.

It is ingenious, said Whitbread smiling happily, even though you will probably charge me an arm and a leg.

It is most effective, I replied, but it will only cost you an arm and it will ensure you grow many more to drink your Tea.

Whitbread laughed, you are the man of the moment Watt. My men cheer you, the Royal Society, I believe cheer you, and the King himself and his Queen came to visit me only to view your marvellous steam contraption, and I hear you had an audience with His Majesty some time ago. Whitbread Ales should be Honoured to have your engines on its premises!

The King's visit was a marvellous Occasion for all concerned, I replied. And the King most gracious. And if it pleases you, Whitbread, sir, I will be about making your engine *a la mode*.

We gave each other a short bow and left in cordial understanding but I was certain my men had built up a thirst. I thought him a mean brewer who could not offer his engineer an ale. I made my way back to the men waiting on the brewing floor and learnt that Whitbread's men had named Boulton & Watt's men Ale Tasters and we would have to satisfy them on the quality of their ales when we had finished our 'tuning.' We should have cheered Whitbread, for we had a fine afternoon's work ahead of us and days after that.

*

57

'I will pay no heed to these Smotherers'

Richard Trevithick is the son of the respected mine manager Richard Trevithick Snr. He was poor student but a skilled and courageous engineer. He is 24 years old and an engine wright at East Stray copper mine near Camborne.

Richard Trevithick
Engine Wright
Camborne, 1795

My father warned me in my early days to keep clear of the Birmingham Smotherers. They will be all over Cornwall, he told me, building this and building that, tuning this and tuning that, and all at our expense. If we lift our heads and think for ourselves we will be branded Pirates and put in the hands of lawyers.

    He would shake his head at my drawings and warn me that I would be spending my days in filthy courtrooms in Chancery Lane if I be bold enough to challenge Boulton & Watt. He has spent his life in the hardships of mines and deserves his rest, so I will not be the one to tell him grubby notices have appeared on our doors declaring us to desist as we are in breach of Watt's Patents. They have taken on Jonathan Hornblower for his excellent double-acting engines, but he was undone when the Courts would not honour his Patent above Watt's. Now Edward Bull and myself face their bullyboy sons with Injunctions in their hands. If I can bear the smell of lawyers, we will take them through the Appeal Courts and make the wheels of justice grind exceedingly slow for we have only a few years before the shackles of Watt's '69 Patent are dissolved. Wm. Murdoch is a Soho man and Watt's Spy among us. It is his burden to swear affidavits against those whom he believes breach of Watt's patent. He has become our Chief Inspector of Engines, an Office gifted to him by our Soho overseers. There are Mine Captains who do not want him about but there are many who are the better for his practical teachings. It may surprise some that the Cornish know a good man no matter the clothes he wears for he has lived and breathed in Redruth long enough to be called a friend. We have shared the life of steam men in all weathers and bent our minds over a few ales to what we can call our own.

He is a noteworthy figure and even my father will give him the time of day and doff his hat to this northerner. It is said he walked from Cumnock in Scotland 300 miles to Birmingham to join James Watt. It is also said that Matthew Boulton the great man himself was mightily impressed by Murdoch's wooden hat made entirely on a Lathe. But what I know is that he is the truest gentleman I have ever met and the most inventive and resourceful.

When he talks of his inventions he does so in such a manner of modesty and practical application that he has me hanging on his every word, or so my father used to teaze me. He talks as if he has a Steam Manuel set right in his mind and need only turn the page!

He explained how he changed the gearing mechanism of the steam valve to so it could be worked from the action of the exhaust shaft. It is well known among steam men that he designed the Sun & Planet Gears which allowed Watt to continue with his rotary engine.

I was only a lad of ten, when my father told me he had seen Murdoch draw his solution to making a vertical motion into a rotary motion in a few simple lines. The 'planet' was a cogwheel that turned around a central cogwheel the 'sun' when driven by the beam of the engine. When the 'sun' was turned by the 'planet' it drove the drive shaft. My father also told me James Watt had patented Murdoch's Sun and Planet gears in his own name, *tout suite*[1].

There is not a Boulton & Watt engine in Cornwall that does not run at its best from Murdoch's care and since they make their fees from fuel saved he is their golden goose. They owe him their fortune in Cornwall and would suffer some mortification were they to hear how much weight we put on Wm.'s skills at their expense. There are stories about him which he shrugs aside as if he is about to blush but this only makes the listener beg for more. It has long been around Redruth that he has lit his house with gas. That any man would be brave enough to do so is astonishing enough but that a man should be able to do so is a mystery that Redruth guards as its secret joy. There would not be a week pass by that one man or another tells the story of Wm.'s lantern. They make it the most devilish black night that stopped his workers from walking home for fear of breaking their necks on our horse tracks. Wm. took to making a lantern. He filled a bagpipe bladder with gas and used a tobacco pipe as a funnel for a thin steam of gas which he burnt to light the way!

For myself, my interest lies more in Wm.'s steam carriage. It is well-known that in '84 he had a model steam carriage run around his living room in Redruth like an iron chicken looking for its head. It had three wheels and was about one foot in height with the engine and boiler between the two larger wheels at the back. A small spirit lamp heated the water and a tiller at the front turned the smaller wheel.

It filled many with horror that such a contraption could be set loose upon our roads but the idea of a man-made machine moving of its own accord filled me with such joy I vowed to make something of my life with such Wonders. Wm. has told several engineers that he plans to make a steam engine which can draw carriages behind it. He was certain that his idea will answer and that there is a great deal of money to be made from it but he has not had the support he wanted from Boulton & Watt.

Last year with my cousin Adrian Vivian I visited Wm. at his house in Redruth, where he was most convivial and happy to demonstrate his steam carriage. We watched it work spellbound with its simplicity and ingenuity. It used strong steam, that is, steam pressure alone with no vacuum. But it was not only a steam carriage which has occupied his mind. He discussed his plans for a unique design for a compact cylinder steam engine which requires no valve gear and would be cheap to make. We rode home to Camborne in the late afternoon with very small conversation being completely taken up with what we had seen. I, for one, was caught in a world of such striving and glory I arrived home with a pounding heart as if I had taken it into my head to run the three miles.

This week we have heard that Wm.'s plans have come to Nothing, being thwarted by Matthew Boulton in person. He was on his way to London to register his Patent when Boulton met him in Exeter and, by some means I can barely comprehend, persuaded him to forgo his Adventure. He came back as far as Truro and, being reduced to a Monkey Grinder with his Monkey, he took his steam carriage for a demonstration run around Rivers Great Room at the Kings Heads Hotel, where he had it carry a fire shovel, poker and tongs. I can barely think on it without my blood running hot enough to make of Boulton a sorry man. I am considered handy at hurling and Boulton will be lucky to keep his feet on the ground if he has the Gall to wander into Cornwall. There is nothing for it but to take on these Birmingham Shagbags and make them pay. I am a lucky man to have David Giddy by my side for he is so rich in the ordering of numbers he will not let my calculations go astray.[2] There is a Brotherhood of steam men counting down the years until 1800, when we will show these Soho Smotherers they are not the only steam men in the world.

*

Ari's Birmingham Gazette
1 February 1796

Two fat sheep (the first fruits of the newly cultivated land of Soho) were sacrificed at the Altar of Vulcan and eaten by the Cyclops in the Great hall of the Temple which is 46 feet wide and 100 feet long. These two great dishes were garnished with rumps and rounds of beef, legs of veal, and gammons of bacon, with innumerable meat pieces and plumb puddings, accompanied with a good band of Marian Music. When dinner was over, the Founder of Soho entered, and consecrated this new branch of it by sprinkling the walls with wine, and then in the name of *Vulcan*, and all the Gods and Goddesses of *Fire* and *Water*, pronounced the name of its SOHO FOUNDARY, and all the people cried amen. These ceremonies being ended, six cannon were discharged, and the band of Music struck up God save the King, which was sung in full chorus by two hundred loyal subjects.

Original Document: *Ari's Birmingham Gazette* 1 February 1796 Opening of the Soho Manufactory

58

'tears to my eyes'

Matthew Boulton inherited a button-making factory. He was both a man of business and an inventor. He has been mass-producing goods in his Soho Manufactory for twenty-eight years. He is 68 years and has opened the Soho Foundry which contains the Soho Mint. He is a Fellow of the Royal Society of London and a Fellow of the Royal Society of Edinburgh.

Matthew Boulton
Soho House
Birmingham, 1796

I will happily post a copy of the Gazette to Watt, for it could not be a better description if I had written it myself! And in reply to daughter Anne's harping on, I have this afternoon taken rest in my study and given myself up to some rich meditations. I will turn a blind ear to the whispers I have heard from the kitchen that my guests took with them as many Knives, Forks, Pepper Boxes, Spoons and Mustard Pots as they could stuff their pockets with. As dear Wedgwood, God rest his soul, would remind me, do not think such thieves are pissing on you for those Trinkets carry your Name and keep it in the minds of many. Fame, Wedgwood used to say, is a most reliable salesman.

 It brought tears to my eyes to hear the cheers of our workers as we celebrated. They sang for Church and King with great fervour given the horrors in France, but their eyes came to me and their loyalty is a steadfast rock for Boulton & Watt. I had six cannons discharge to honour Soho but every man Jack of them knows I will arm them against any Tory mobs rampaging near Soho and no one will threaten their livelihoods, or their lives, while I have breath in my body and cannons by my side!

I was lucky I did not start to weep in publick for I will confess I did return to the house a weeping man and sat for a while with my heads in my hands like a boy with a lost dog. I have had the greatest satisfaction of late and would be thought a happy man yet momentarily, success seemed a poor consolation for our endeavours, and I felt as I had done a thousand times, a man skating on thin ice, ready for a chilly death in the arms of a Banker.

This afternoon I have revived my spirits and counted our success as the Blessings of good labour. I have been on my knees so many times I have worn a grove in the floor – but I look around Soho and I see we have grown and flowered and reformed as Darwin would have his animals and plants and there is not a company in the land who can match us! We now control production from plans to assembly and will make our engines entire, given their own frame and moveable to be sold complete from our Soho Works! It would be a long list to name all the cotton, corn, linen, silk and brewery mills that boast Boulton & Watt engines. We will have the rope-makers, dyers and dockyards next on our doorstep wanting our business!

I should count too, as my great Blessing, that my Mint at Soho has, at long last, stopped draining my pockets and produced what I will say are the first of their kind. My coins are fully round, mass produced and steam powered! Furthermore, it has given me great satisfaction that our own Company foundry will make us no longer at the mercy of the Wilkinsons at Coalbrookdale, who think their monopoly on quality will never be broken except they break it themselves, which they seem bent on doing with their family quarrels.

Which brings to mind James Keir has lately chided me for an Intrusion he claims belongs to me and which he describes as a great Dilemma brought on Earth by my manufactory. He said he could not remember a time when seconds and minutes were so important and the blame settles with myself and Watt.

You have made us count production by the hour and an engine by its revolutions and Time, once so benign and predictable, has gone forever, he declared on his last visit.

What then is your Dilemma, I asked for I would not let him play me so.

My Dilemma is that I have learnt to love my pocket watch, carry it with me everywhere and am quite lost without it, he replied, more than half serious.

Then you must blame Emery at Charing Cross for bringing his Swiss ingenuity to London, I parried.

You are probably right Boulton, he replied, but if Emery had never crossed the Channel to make us better pocket watches, you would have Murdoch or Watt see to it.

Since your alkali production at Tipton has made you Soap-Maker to the Midlands, I countered, you should be glad to know the time of day! We laughed, well pleased with the lie of our land. It has become a general belief that Watt and Murdoch can do anything they put their minds to.

I told Watt many years ago I was after making engines for all the world, but I must confess I was not certain of the rock sitting before me. If ever a man was reluctantly dragged from the clutches of Obscurity it was James Watt. It would be another long list to name Watt's nurses through his Maladies beginning with dear Small, Darwin, Black, Roebuck and myself, but for all that, he had been a true and steadfast engineer.

It may ruffle Watt's feathers to hear that I place William Murdoch as another rock upon which my whole Edifice rests. Keir is right to say that his ingenuity keeps pace with Watt's. As my manager, he will turn his wise-eye upon our quality, for I will not be a man who is known for shoddy masses of goods the world will only want to shit upon.

There are new rocks for future building. We plan to become Boulton, Watt & Sons with my son Matt, Jim Watt, and if he is so interested, his brother Gregory, the young geologist, as part of the company. As for myself, it has not made me weep to count profit, long a stranger to my ramshackle balance sheet. There is now more than enough of it to make me need a book-keeper. Since I lack the will to master such things, I will find a man who can, and the name of Boulton will not be sullied by jibes about what matters most in business: the proper keeping of accounts. I should thank Keir for noting when he was our Manager that I ran the enterprise like a captain plotting a future course while the ship drifted where it might. Those were the days when our debts mounted and men stared at me in astonishment when I predicted there was money to be had in steam engines.

As I write, I have set about putting our young partners onto claiming their proper royalties and bringing to brook these engine Pirates wherever they may hide, including the rocky caves of Cornwall. Of all the victories Watt and I claim, none will be greater than to see our sons stand beside us in business, so quick and learned and fresh to the world, looking to the future as Soho will always do.

***

REVOLUTIONS

PART 2

Steam Team 4

Matthew Murray 1765 -1826

Richard Trevithick 1771 -1833

William Hedley 1779 -1843

George Stephenson 1781 -1848

Robert Stephenson 1803 -1859

59

'She was going faster than any man could walk'

Richard Trevithick
Christmas Day
Camborne, 1801

If a man cannot laugh and have a few ales with his mates the world can go be dam'd with its cant and claptrap. I am said to be an obstinate man but I am surely a sorry one today, for the drink has put a thousand hammers to my head. Inn-keeper Marrack is no stiff-rump but he asked for 'board' acquired when the four of us slept in his dining room. He was grateful for small mercies he said, seeing we had not pissed on his tables or napped with his coaching horses, which he valued well above ourselves. Still it is a fine Christmas for we have made our little history story and will laugh till next Christmas.

My steam chariot was put into shape at a Smithy's near Redruth and I asked the boys about if they would be in the lark with me. They named her *Puffing Devil* for her steam puffs that rose like white clouds. She had a pressure-operated piston connected to a cylindrical horizontal boiler and I aimed to get 145psi out of her which is one in the eye for the Birmingham boys who quake at anything over 25 psi!

I was a mind to take her up the Weith to Camborne Beacon. It was a stiffish climb but I trusted the boiler with my life for it was made by John Harvey's men at his Foundry. My wife boasts there isn't a better one in all of Cornwall and, since he is my father-in-law, I readily agree. It was my cousin Adrian Vivian, already a steam man, who insisted he be at the controls. About six of us went for I promised them a Christmas Eve Ale in Camborne.

She got up steam in the road outside, and we jumped on, that is, all six of us, and I told the boys to hang on tight, which they did rather silent until we got to the top of the hill, and then she went like a bird and they shouted hoorah for Captain Dick. After about a quarter of a mile, she met loose stones which did not suit her wheels and slowed her down, then the blasted rain frightened some of the boys who jumped off in case she bucked them off. It pleased me that she went faster than any man can walk for another quarter mile, where we turned her around and she went good as gold back down the hill. True to my promise we went to Tyacks Inn and that is why I am a useless engineer this Christmas.

I will tell anyone who will listen after this day that if I am the reckless bastard they say I am, I will prove them all correct in principle and in practice for I will make steam my servant, not my master, and you can give that Mouthful to the boys in Birmingham.

Richard Trevithick
Puffing Devil 1801

It is three days after Christmas I am getting roasted around drinking tables for the fate of the *Puffing Devil*. She now lies in a ditch. Harvey's men are mightily unimpressed and tell me my men left the fire still burning when they "broke" the Devil and took themselves off to the nearest public house to feast, I am reliably told, on goose and ale. Such engines will not tolerate their water boiling away and leaving the fire to do its worst. No wonder she "blew up." I have learnt one certain thing, without reliable men we will not have reliable steam! They can kiss my Arse before they touch another of my engines!

*

60

'I will not let my world be as small'

George Stephenson has recently been appointed brakesman at Killingworth Colliery, Durham. Illiterate until he was 18, Stephenson is now 22 years old, a self-motivated, talented mechanic interested in, and fascinated by, the winding steam engine he tends at Killingworth.

George Stephenson
Lime Road Cottage
Killingworth, 1804

My mother, so my father tells me, would have been mighty pleased to be a grandmother. And my father adds, in his usual way, no matter how old a farm girl is, she will pop them good and proper! My father thought to share a laugh but I did not take to such jesting. Fanny Henderson may be a farm girl and a woman with years upon me but she is the sweetest soul I ever met. It is wonderful news that she is with child again and we will add to our family until we are a snug fit at Lime Road Cottage for which I am lucky enough to have the keys in my pocket.

Robert is sitting up and cooing at the world as if he is most satisfied with it and that makes me laugh. I sit him on my knee and we mend shoes together and I wonder what it is he would like to learn although Fanny says he is such a strong child he will have no trouble making up his own mind. Whatever he decides I will not have him down at pit at Ten not even to satisfy my father who says a boy learns a lot down the pit even tho' it is many years since he has been down one himself.

I will not have you working for crumbs, Robbie boy, I said most seriously to him, No I will not. Whereupon he raised his little fist and waving it about had me a good one right to the eye!

Fanny and Robert are my chief delight but it is learning that takes up my mind, from the light to the dark, and about all that is front of me. I will not let my world be as small as a pit, or a shoe-maker or a man to service a winding wheel, for masters who know less than him. When my mind is free from work I consider many fancies. There is one which is a constant puzzle. It is a Perpetual Motion Machine, which my teacher advised may 'violate a law' from the great Philosopher Isaac Newton, for it seeks to create energy out of nothing which I do not believe it does at all, being instead a perpetual depository for the holding of energy once started. I turn this over in my mind as I go about my day for it seems a most certain thing. There is Bishop Horsley's publication of Newton's work in English (being unschooled, Latin is well beyond me and everyone in my family since time began) but my teacher says it is well beyond his and my funds. He suggests I think on why Newton would say momentum, energy and angular momentum cannot be created or destroyed.

I do know this. I had cause to mend our eight-day clock some time ago it being most badly soot affected after our fire. By observation, piece by pierce, and relying on the cleaning of parts, I achieved a happy chime and it has worked since. It occurs to me that a man can work by observation and examination not only from the study of Latin books. If I was to confess any interest in things possibly, indeed, quite probably beyond me, I would confess to a carriage or some such, driven by steam and able to carry my companions, or other loads now done by horses.

It is known that the Cornish mechanic, Trevithick, who people say has steam in his blood, has made such a chariot and has it on rails in Abercynon, which I am told is in Wales many miles hence. It is also said he is a fearless madman …….

*

61

'Homfray was waving ribbons'

Richard Trevithick
Merthyr Tydfil, 1804

The Welsh Ironmasters have given me a fine steam circus with their rivalry. They have made steam all the rage hereabouts and every man and his dog is talking about cylinders and flues. There are crowds about everywhere and they sing and cheer for the Welsh are hearty fellows and don't mind a bit of noise.

There was a great Wager afoot and the local villagers were stirred to place bets themselves and sang my praises before I'd earnt them. Samuel Homfray who now owns my Tram Engine, (but not my Patent) started the Grand Affray. He wagered Richard Crawshay from Cyfarthfa Ironworks 500 guineas that his engine would haul 10 tons of iron along the Merthyr Tydfil tram road from Penydarren to Abercynon. I am not too proud to take a challenge to make my engines known.

The occasion stirred so much interest I had a hundred cheering witnesses. They cheered the hissing and thundering and snorting of steam and cheered the men who boarded her, either with envy or admiration, then they cheered Homfray who cheered back and all were ready for some sport never before known!

My Pen-y-Darren engine is as strong as I could make her. I placed my high-pressure cylinder and a long piston rod which drives an 8 foot flywheel on an iron tray. Strong steam requires strong design and my

Pen-y-Darren engine carrys a new Calculation. She has a double-acting cylinder with a four-way valve to distribute the steam. High-pressure steam has no need of a condenser. I have turned the exhaust steam up the chimney to create a draught to draw hot gases from the fire more quickly through the boiler. David Giddy has done the calculations for me and assures me that such a design will hold true. This works so well it will set the future for such engines, though the crowd that gathered cared less about such particulars and more about the steam smoke she blew.

We started in great style with noise aplenty, the crowd outdoing the engine and Homfray leading the charge. She went like a beauty and pulled ten tons of iron in five wagons, and 70 men riding on them for the whole nine miles. The engine pulled at between 2-5 miles per hour and the boiler had no need of water on the smooth level run. The coal consumed was two hundredweight.

Homfray was waving ribbons and claiming victory and calling his drummers to beat out the same. Crawshay was much impressed but would not settle until the wager had been completed with the return journey. This I went about with great confidence until a bolt snapped causing the boiler to leak and we did not get back until the following day. This gave Crawshay reason to claim that the run had not been completed as stipulated in the wager, and I thankfully left it to the gentlemen involved to settle their dispute.

Despite my success and the beating of drums, I have many detractors. James Watt himself has made it known I deserve hanging for using high pressure steam. Boulton & Watt have been speaking high against me since the disaster at Greenwich, for they will do anything to spoil their competition. They will not bless a future which boasts a high-pressure cylinder without a condenser and I should thank them for it, for I will outdo Boulton & Watt and double, nay, treble, their working power. Their engines may be mighty by reputation, but they have never moved an inch from their blocks and challenged horses that pull tons of coal! Meanwhile they seek to blacken my name and call me heartless.

There is no one who does not know the explosion of my pumping engine at Greenwich which killed four has brought me great consternation. I was bound by honour to inspect the scene and it tore at my soul. The boiler lay split and twisted and I could not help but note the blood of good men upon iron. I was told it blew the fireman 100 yards and left him so his own mother would not know him. But the engineer is required to find the fault and I concluded the design failed to accommodate what I called the Operating Factor, for the fault lay with the boiler man and not the boiler. I cannot leave an engine at the mercy of its operators, but neither can I leave men at the mercy of the engine.

For this purpose, I have included two safety valves in my new engines; I placed a disk to cover a small hole in the boiler above the water level in the steam chest. I attached a weight to a pivot lever to equal the force of the steam pressure. By adjusting the weight on the lever, the operator can set the pressure. To add to the safety of the boiler, I have placed a simple lead plug just below the minimum safe water level. If the water runs low, the temperature will rise and melt the lead plug and release steam. Happily, this will reduce pressure in the boiler and the whistle of escaping steam will warn the operator. As chemists have claimed useful expansion properties for mercury, I plan an additional device, a mercury manometer will measure pressure and add to the safety of the whole.

I am happily planning to visit Matthew Murray the new steam genius at Leeds, and view his giant Steam Hall in Water Lane, where, the world knows, he intends to produce all manner of steam engines which will, in time, challenge Boulton & Watt. They may laugh at me but they do not laugh at Matthew Murray. They are in the courts defending what they say are Murray's infringements on their patents and they have, I am reliably informed, bought land in Water Lane to stifle any expansion of Murray's works! Murray shrugs them off and moves ahead as I do!

As for the present, my trials continue. My Pen-y-Darren engine runs on iron plates designed for the lighter axle load of horse-drawn wagons. I thought she would manage it but, when the plates broke from time to time, the engine fell mightily out of favour. Homfray, who declared himself exhausted by the whole proceedings, cursed my engine as a great bothersome thing. He complained that it would never reduce his transport costs and had it dismantled and set up in its original station to drive hammers. I cannot have delays, he told me, or Parliament will be on my back. I did not blame him for we are at war again with the French who should have learnt their lesson last time we thrashed them which we will surely do again!

I returned to Wales this summer to find horses once more working the tramway and the engine driving Homfray's hammers and the world around them happy about its business. But whatever the world may decide, there are engineers aplenty, not given to the lures of agreeable distractions, but duty bound to find a place for 'strong steam' worthy of its great power!

*

62

'the purpose of my visit was to show you this'

Andrew Cosgrove
Handsworth, 1806

I have returned from Heathfield Hall and James Watt Snr., as he is known now, his son James Watt Jr. having become a partner and manager of the Soho Foundry and Manufactory. Josiah was waiting with his dogs, Tilly and Angus. He had set us some ale and comfortable stools by Amy's hearth fire.

It is many years since I have seen Watt and many years since I have held a pen and sat down to write my thoughts. My inspiration, once so hot upon my mind, turned to ash after my dear wife Amy's death from a sudden winter chill no doctor could placate. After the local apothecary suggested a bone broth of ground human bones at great expense I wanted to call on Doctor Darwin but we did not have the funds to bring him from Derby.

Josiah did all he could for both of us. I am ashamed to say I was so taken by my grief I was a burden to be carried, rather than a father to be leant upon. I have made him a partner in my business, such as it is, for he is not only the best blacksmith I have known, he has been a son to me and I a father to him.

In response to my letter of enquiry Watt informed me that he had retired from his business some years ago and was doing further experiments in his workshop, but would be glad to pause for an afternoon of my good company. My invitation was to Heathfield Hall, a most elegant house befitting a man of Watt's fame and fortune. A gardener showed me the way to his garret workshop up the top of some narrow stairs which give it a most private air.

I thought myself an old man as Watt appeared. He had a sprightly step and held out his hand in good humour and bade me enter what he called his Cavern atop the Trees. If I understood the drift of the house it was in the roof atop the servant's rooms. It was a small room with plain white-washed walls lit by a low window looking into shrubbery but dull for all that. There were so many bottles, casks, containers and jugs packed on shelves I could hardly make sense of it. I did recognise a foot lathe against the window but near it was a type of Saw I had never seen before. I stood like a man bemused and took it all in. There were parts of flutes and violins lying as if suddenly discarded. He had made, it seemed to me, a cosy nest for himself and to keep him company, his snuffbox, a fruit plate and his clay pipe. In the corner was a small stove where he could heat his meals. His wash-leather apron lay adrift across a chair as if it had been hastily discarded by its wearer on hearing footsteps on the stairs.

He had put on his coat to meet me as the gentleman he had become and pulled up a chair for me muttering that it had been years since we had shared our thoughts. I am certain he was set for a convivial conversation with his old blacksmith but I had climbed his stairs with a burden so heavy I had to release it.

Mister Watt, sir, the purpose of my visit was to show you this, I said, and put Tom's medal in its silk kerchief on Watt's work bench.

And what is this, Watt said, leaning forward to touch the folded silk with some curiosity.

It is Tomas Winterson's Trafalgar Medal, I said and unfolded the silk.

It was copper with a silver sheen. It had a royal blue ribbon threaded through its loop.

Tomas Winterson? Watt looked puzzled. I do know that name from somewhere. Smithy Cosgrove, please jog my memory.

It is some twenty years ago I asked for your help through the good Doctor Darwin to vouch for my apprentice, Tomas Winterson, in his quest to become a Midshipman in the King's Navy, I explained.

Watt looked concerned. I have no memory of such a request, although I do not doubt your word. I am sure the dear Doctor would have responded favourably, then he paused and said, you know Erasmus passed to the Good Lord some four years ago?

Yes sir, his death was a blow to all in the district, especially Lichfield, who has such fond memories of him, I replied.

He was a dear friend to me, said Watt, and idly fingered Tom's medal. What is his story, he asked, by which I took him to mean Tom's story.

Because of letters he had collected and especially from the esteemed Doctor, he was successful in being taken on as first a servant to officers and then a Midshipman, I said.

He became an officer then, Watt enquired.

First Lieutenant Tomas Winterson served on the *Temeraire* under Captain Eliab Harvey and was wounded in action at Trafalgar, I informed him.

Did he, said Watt, his face flushed with his feeling. The fighting *Temeraire* helped save the *Victory* I believe, although nothing could save Nelson.

Indeed, I said and I could not hold back the bitterness and added, nothing in the end could save our Tom. He died of his wounds this last month and leaves a wife and three children.

That is terrible news, said Watt. The whole country is in mourning for Nelson and our brave heroes. We are all, every man and woman of us, eternally grateful.

He held the medal and read out the words engraved. Tom's details were in the top quarter of one side showing a bust of Lord Nelson and on the reverse a scene of the battle with the words, ENGLAND EXPECTS EVERY MAN WILL DO HIS DUTY. In the reverse exergue was TRAFALGAR OCTr. 21 1805.

I see Boulton has left no one in doubt as to his commemoration, Watt said, and read round the edge.

From M. Boulton to the heroes of Trafalgar.

It is a well-meant gesture and much appreciated by the families, I replied.

Watt looked at me directly. And Cosgrove, he asked, do you want some funds to help your Tom's widow and children? I'm sure Doctor Darwin would have me ask. He was always of the belief that war was a sign of the insanity of human beings.

I knew he would ask and I knew I would say no, for we are never beggars, but it was none the easier in the doing of it.

Are you sure, Cosgrove, Watt replied, and added that he would happily support Tom's poor wife.

Thank you, sir, I replied, but she is a navy wife and most capable and her family is well-connected. I came only at the request of Tom's brother Josiah, once my apprentice, and now a partner in my business. He asks that you convey his appreciation to Mister Boulton for his brother's medal. After your recommendation on my behalf to Doctor Darwin I wanted you to know how well Tom served his country.

Then thank you, Cosgrove, I am most touched. Of course, Boulton would meet this hero's brother to pay his respects. Most gladly he would.

I expect he would, for the whole country yearns to honour our brave sailors, but I would rather Josiah not meet Boulton, I replied.

And why would that be?

Because Josiah is the most able Smithy I have ever known and I would not want Boulton to press him into his service.

Press him?

Yes. Or some such thing, as is Boulton's way. I would not want to lose Josiah because he is like a son to me.

Watt was silent for some time. When he did speak he looked drawn and I watched his energy slip like an ebb tide.

I know what you feel, Cosgrove. I recently lost my son Gregory to consumption. I would be honoured to inform Boulton of the appreciation of the Winterson family.

We stood together in front of the medal, both of us very still.

I feel it every morning as I rise for the day, and every evening as I slip so quickly into sleep, I am fading, but I knew my hands were still strong enough to snap Watt's bones in half.

I was shocked to know my thoughts and turned my head away, even as Watt was being gracious.

The question came to me and I did not stop myself.

Do you ever hear your engines speak, I asked Watt.

It was a question without prudence or manners and it left me standing like a man naked before the world.

Speak?

Yes. Talk to you.

God, man, Watt said and laughed. My engines, according to those who know me best, have been muttering to me for years. Each one full of gripes and fancies and usually no gratitude for all the care I lavished on them. He laughed again.

Why do you ask Smithy Cosgrove, with such a solemn look upon your face?

It does not signify, I said and I spoke the truth. What can one man's imaginings signify in the face of a Revolution? I was the smallest man alive amongst giant waves and wholly unable to repulse them.

If you have heard my engines speak I hope they were more civil to a blacksmith than to an engineer. Watt laughed again, his spirits returning.

And Smithy Cosgrove, I seem to recall you like cider. Would you join me in a tankard?

He poured me apple cider and showed me, step by step, his workshop. He spoke very fondly to me as if he valued my consideration. He showed me a machine he said would standardise the size of coins for the first time, and was proposed for the Royal Mint. He opened his drawers and revealed his many odds and bods of screws, punches, taps and dies as if he thought we might need one such item any moment. He showed me a curious circular Saw and the frame of a machine for copying Sculptures in miniature. Without thinking it may be beyond my understanding he explained its principles of operation and I saw how he had thought to make it use two hands. One hand would travel over the pattern and the other would cut the material into the form of the face or the shoulders.

I commented on the red rag covering some wool around a beam and he laughed and admitted it was where he was prone to hit his head.

You would be a danger round a forge, I heard myself say, somewhat in horror at my honesty.

Watt nodded. Then he told me that the public only saw his success whereas he remembered his many failures. If I have excelled, he said, more to his cider than to me, I think it has been by chance, and by the neglects of others.

We sat and shared our cider and I felt as I had many times before, a great admiration for him. I did not ask him if he missed making steam machines because I knew he did not. He was no longer a man rolling a great stone up a mountain. He was a bright-eyed bird happy in his nest.

*

It is now late and a wind howls but I am warm and fed and Tilly is at my feet. Yet I cannot sleep and write my tumbling thoughts. It is a long time since Amy Cosgrove, nee Newton of Leicester, brought me in my supper and a longer time too, since I first met Watt coming down the hill, rubbing his hands with excitement as if he had come upon some golden treasure. Now there is treasure aplenty for the men with engines and plenty of others who swear by them. Steam feeds the Manufactories and the Manufactories feed steam, and all about hungry women and children flock to them and say Yes sir, No sir, to new Masters. These are the times we live in.

Maybe Amy would quote me *Ecclesiastes* and remind me there is a time for every purpose under heaven so I might reconcile myself with the world. And I would probably argue that *Ecclesiastes* was a very pretty list that any philosopher could draw who had not seen the new Heaven.

She might be surprised that I would question so, but our new age is full of questions.

But one verse that *Ecclesiastes* speaks I know to be true.

> *That which befalleth the sons of men befalleth beasts; as the one dieth, so dieth the other; yea, they have all one breath; so that a man hath no preeminence above a beast: for all is vanity.*

If I write more I fear I will begin weeping, so I will put down my pen and pray that nothing brings me to see whatever Hell cometh and how brazen it may be.

*

63

Catch-me-who-can

Richard Trevithick
London, 1808

I would say to my critics *Catch-me-who-can!* But that is a laugh for no man can if they observe the engine I have this summer brought to working order. Whether it be sunny or a London drizzle, my railway runs around a track in Euston Square on which *Catch-me-who-can* moves for the pleasure of the public who have a shilling to spare. She was named by Giddy's daughter who clapped her hands and cried for all to hear, 'faster, faster!'

I mean to have the populace enjoy my contraption see its great advantages in transport. I have called my track a 'Steam Circus' and have made a fine start at reaching 12 mph much to the delight of my passengers. I have also invited many adventurous gentlemen to see my locomotive in hope of raising funds for further designs but I must confess I am not without trouble.

Catch-me-who-can is a different design to my other engines and was made by Hazeldine & Co at their Bridgenorth Foundry for I cannot afford bolts to snap, rivets to break or seams to burst. John Rastrick promised to see to its casting and I had confidence in his results. I know him from working with the Thames Archway Company on its tunnel under London's great river. I will not have it put about that steam has reached its end as our tunnel did when we could go no further. Steam engineers are well on the way to great success unlike our miners who presently lack the methods to tunnel under such rivers. Men with picks in candlelight cannot progress against the horror of its quicksand, silt and ancient debris so the river waters seem as deep as the earth itself.

Catch-me-who-can is as neat as a pin and I am very proud of her. She has no huge flywheel, horizontal cylinder or gearing, but rather a vertical cylinder encased within a boiler that drives one pair of wheels by connecting rods. I have kept the boiler as a return-flue type with an internal firebox., that is I have, by way of steel tubes, placed the heat within the water boiler and not under it. But for all my improvements, the cast-iron rails are too brittle for the weight of the engine. It only took a few weeks before my fears proved true. *Catch-me-who-can* broke the rail bearing her forward and tipped heavily on her side. I retired to the nearest public house and made merry instead of cursing all about me.

A man wants to be known for his achievements, even if mine are to have an engine that breaks rails. Any Judge of fair mind would say I have parted the pages of history so the locomotive can drive through to a glorious future. Others will use my designs for the return-flue boiler and the steam jet through the chimney, for they are the most efficient design possible.

That I can find no investors of like mind will be answered in time. Our strong steam engines will prove their worth because they are smaller and less expensive to make than Watt's lumbering giants. This will prove a fundamental truth. They will, by and by, become faster and faster. They will have a thousand different uses, and I will list them by and by. Meanwhile, I am a Trevithick and I will be dam'd if any man will sway me from my purpose. And I will drink to that and go tonight to Pall Mall and marvel at the public gas-lamps the Saxon Friedrich Winzer has given London. I will curse the Dullards who welcomed his ideas so much, he now lives in Paris.

*

64

'a lumbering puffing iron cart!'

George Stephenson
Lime Road Cottage
Killingworth, 1812

It seems I must be forever writing. I will not be a man measured by his schooling, for there is little of that to measure. It is a satisfaction to me that I have men about me who have not seen the inside of a grammar school. They would not know Latin or Greek if it called them Father, and I am pleased to call them friend. Yet there are letters that need answering and inquiries concerning our plans at Killingworth.

Today I have something worth the telling. I have returned from Newcastle town, in particular, Coxlodge colliery, where I am told the future lies. Very early this morning Nicholas Wood came to me with a distracted look about him and said I should borrow his horse and find out how History was being made almost on our doorstep and we had no inkling of it. He did not need to offer twice as I could be back before dark and avoid the nuisance of boggy roads across the moor.

We had already heard that Matthew Murray has advanced Trevithick's faith in pressure steam and has used a twin-cylinder design to create a strong engine. More to my interest he had placed his pistons in a horizontal position which seemed to me to be a likely benefit. Wood saw me off and shouted after me that John Blenkinsop, Middleton's manager, was claiming that their *Salamanca* engine has made history and was pulling coal to the straithes(1) faster than any horses on earth. Moreover, Murray is letting it be known it is more than paying its way in savings! Wood shouted so well I think he expected me to gallop the whole six miles! If rumour be true then Matthew Murray will have more colliery managers than Nicholas Wood lift their heads and do their sums.

I arrived with great hopes of meeting Murray himself but his men told me he was in Leeds where he has much business. I was not left to cool my heels for his men were eager to have me marvel at *Salamanca* which they impressed upon me was named after Arthur Wellington's latest victory. I doubted if any of us could point to Salamanca on a map but we knew the Spanish could never trouble an English army in battle and the Duke was a new hero to match Sir Drake at his Bowls. We were in fine humour and Murray's men discussed his engine with such open hearts I was at first concerned that Murray might not, if he was present, be so forthcoming.

As they stood around me, their caps in hand and their faces weathered with their work I saw that they were not miners but steam men; men of a certain type, that is, they were fussy men with an eye for detail and a love of precision. They were so certain that *Salamanca* could not be bettered they saw no harm in conversing with a visiting engineer. Besides, they told me that in Murray's workshop they were about building the best engines in the Kingdom, not warring with other firms defending patents!

I do not know what they thought of me but they looked me in the eye and told me that Murray was the greatest inventor in these parts, not only in locomotives but in the flax mills as well. I nodded as much as I could but my spirits sank as I contemplated Murray's work. Who was I, I thought, to be standing in the muddy yard at Coxlodge with a smile on my face and a spy cap on my head!

They told me that the engine works at Middleton Railway, along the old wagonway to Leeds harbour, where it pulls eight wagons of coal. They laughed and added Murray himself was still in awe of the fifty brave souls who climbed aboard on its first run.

It is clear he has taken notice of Trevithick's *Catch-me-who-can* but had made his engine more than a toy for penny rides. I was told the engine weighs 5 tons and builds up a pressure of 25 psi. I saw for myself that Murray had embedded two cylinders (I was informed they are 8"x 20" and most reliable) into the top of the centre-flue boiler which drive the pinions through cranks.

I walked around *Salamanca* and saw he had done away with the usual parallel motion linkage and used sliding guides for the piston crossheads instead. I had no trouble with this as it enables a longer stroke and it clear that the length of the stroke allows the engine to burn a greater quantity of fuel and so develop more power.

There is no surprise in the usefulness of a crank shaft to convert a piston's motion to rotary motion but Murray has thought to place them at right angles, and so were my thoughts. His Fireman saw my gaze and was quick to tell me this caused the engine to start wherever it came to rest. I enquired as to the wear on their rack and pinion rails, which was a great curiosity to me, for I had never seen one in use. John Blenkinsop, himself, had designed them and no man of his would speak against them and I was foolish to ask.

I was reckoning to myself that if *Salamanca* weighed 5 tons and pulled 8 wagon loads of coal there would be a weight of 90 tons upon Blenkinsop's rails and I doubted any meshing system would stand the strain. It was also clear to me that Murray's engines pull best when on a level wagonway and that may suffice for Middleton but it will not suffice for Killingworth.

I left a courteous note for Murray and was taking my leave when a dusty old codger approached me. I admit to a certain trepidation but the old wag pulled me aside and said that Murray's Marvel had them all agog round these parts but there is a story of a young boy worth telling. Before I knew it, he had me sat in the waning afternoon awaiting his telling. Apparently, a lively boy lived near to Middleton Pits which he could see from over his back fence. He begged his nurse to take him to see the new steam engine. The young lad waited with his nurse in great anticipation for they were told to watch for a flash of lightning when

Salamanca passed by, and when they heard the engine pulling strong they drew back in fearful joy but all they saw as *Salamanca* passed was a lumbering puffing iron cart!

The wag slapped his thigh and thought it a great story. He made me laugh at his heresy so I gave him a penny for his trouble and told him to find a warm spot for a pint or two, for indeed he looked a man in need of a nourishing ale. I am home partly inspired and partly entertained. Murray has made great advances and given steam legs, but it is not enough. Steam must be made to gallop if we are to claim the wagonways!

*

65

'the beauty of washed linen'

James Watt
Honorary Doctor of Letters Glasgow University
Foreign Member of the Academie des Sciences Paris

Heathfield Hall
Handsworth, 1816

There are hours when I raise my head and cannot believe the chime of the clock or the shadows on the wall as proof of its proclamations. I have bent my body for so long over this or that plan, it is difficult to raise it and account for time.

We are home from Glasgow and Greenock where I visited old haunts and was welcomed by Ann's family, the McGrigors. The world came to meet us, full of marvels. Ann persuaded me to take Henry Bell's paddle steamer from Greenock Port to Rothesay.[1] It is called a Return Trip and most pleasantly filled the day, leaving me much to ponder. I stood on the deck and marvelled at the surge of steam beneath my feet from John Robertson's engine. The *Comet* pulled her weight and ours at a sprightly five miles an hour against a head wind. Ann seemed to find my face the most entertaining delight for she told me I had as many expressions as there were clouds in the sky. We made such good time, if it not for the evidence of my pocket watch, I would not have believed possible. Bell must think me a crusty old curmudgeon for my parsimonious enthusiasm for his dream, for he has well and truly proved me wrong.

To visit where I spent my youth has stirred memories like fish floating from the bottom of a pond and gliding about uncertain of their location. I returned to my workshop thankful the world no longer batters my brain and I go about my hobbies undisturbed, yet today I am full of musings. I have considered washing, that is, linen for bed sheets and so forth. My first wife, so sadly lost, spoke of the beauty of washed linen as if it was worthy of a Poet's glance, laid out all gleaming and holding the smell of the Sun and Wind, whereas I would notice not one jolt of the joy of bleaching linen. That I think on it today maybe the force of another William in my ken, the poet Mr Wordsworth, a friend of Jim, who has settled in Grasmere and writes of his admiration which I cannot, for the life of me, understand a poet would consider, yet here I am considering beauty in the most ordinary of things as Mr. Wordsworth tells us we should do!

I have done nothing but muse on my walks past the Saltmarket and through the Green when I was the proud husband of Peggy Miller and father to a son. It was my habit to be immersed in the problems before me, problems that would mount one upon the other as high as Bills wanting payment, but it is the washing on the grass I have mused upon for more than half my day. And later I mused upon the lime trees from the Physick Garden amongst all the fruit that grew there, for no reason I can think of except that a mulch of lime bark was given to many workers to heal burns which did worry me from the occasions when men from our foundry begged for some relief. It has been many years since I have considered the burns of men smiting my designs. It occurs to me that there would have been no Holland linen spread across the tender grass when I took exercise on that summer Sabbath. It seems so close I can taste the afternoon air.

I envy that young man for his Rapture at his solution, and his unholy impatience with the Sabbath which held back his start upon its Assembly. It was John Anderson, God rest his Soul, who led me to the walk and to the Rapture. He was 'Jolly Jack Phosphorus' to all and sundry for his natural urge to argument and the students' belief he would surely glow in the dark! But I will forever see Anderson in his gown, very smart as a newly minted professor, standing by the Newcomen model he gave me to repair.

Do you think you might do better?

I knew at once he meant I could do better than Newcomen.

Yes, he continued as if I had answered, I glean it is not entirely the fault of Jonathan Sisson that he could not repair the engine. Perhaps the caw-handedness you speak of is Newcomen and not Sisson? I have heard it has proved to be a monster with an appetite for coal beyond the bearing of Industry.

It is entirely inefficient, I recall saying, undaunted by any experience of what I spoke.

Indeed Watt, he then said, I think it is entirely up to the University to prove ourselves useful to Industry, do you not agree?

All I could do was nod that I did.

For what is it to know something unless there is a use for it?

I nodded again.

He rose, tapped me lightly on the shoulder and took his leave.

I remember I prayed he did not expect a solution within the week. I sat numb like a man contemplating a hanging. How was I to do better with an engine that had worked for half a century without any man finding an alternative? I told Peggy of Anderson's challenge and she thought it wonderful. She clasped my hands and holding them to her lips exclaimed that of all the men she knew, I was the one who would find a way. And a way came to me walking that Sabbath on the Green with the breeze fresh to my face. The answer appeared self-evident and I am amazed, even now, as I recall its simplicity.

My life spun on that moment, tho' I did not know it. I would add a separate vessel as a condenser and this would allow the cylinder to work constantly since there would be no need to wait for its cooling. I knew that if the laws of nature held true, my idea would work. The laws of Nature have held true these past fifty years and given a prosperity to many, more than any man might dream.

Jim now complains of our competitors, in particular, Fenton, Murray & Wood at Holbeck, and he grumbles they are precision mad and bent on undercutting our company. He rants about Murray's superior planing machine used for the face of slide valves and thinks it churlish of Murray to keep it under lock and key!

Since Boulton & Watt would not be found innocent of spying I think Murray wise, and have to hide my smile. I cannot judge Jim for his barking at our rivals for he proves every day to be a son of mine. But I do not share with him my secret pleasure at the ingenuity of fellow engineers. I hear that the Richard Trevithick is as stubborn as his father and is well about creating mayhem and will not forsake his grand ideas. It is a happy decision I leave the fray to our young cubs and retire to the peace of my workshop above the spring trees.

Perhaps I will become a poet for I have sat through this late afternoon with the pleasant melancholy of memories past, not of the hard casing of countless cylinders, but the softness of linen prepared by the hands of my first wife and the washing that lay on the Green to bleach by the whimsy of the sun!

*

The world has knocked on my door several times this month. It often does so and whatever I remark upon, it chats the afternoon away about all manner of pressing details. I am consulted about the topography of Cornwall which I am supposed to have memorised, the properties of gases or developments in latent heat or any information I may have regarding copal varnish. It seems I have become the new Smeaton whether I am large enough for his shoes or not. This week I have been asked by the compilers at *The Encyclopaedia Britannica* if I would, if I had time, edit dear Robison's *Dissertation on Steam Engines*.

I have mused over them as well. Dear Robison died in the year of Trafalgar, but before the nation's great victory. He would have cheered our triumph over the French and declared it a just punishment for their excesses. He had the most probing and lucid mind of anyone I have known. I should wait for a round white moon before I review his words for he told me on such a night I had the greater intelligence!

And lately, the Glasgow Waterworks Company writes, asking the whereabouts of my 'sketches' for their water pipes across the Clyde; they remind me it was in '12. They say it was such a successful yet novel stratagem the design should be preserved for Posterity. I recall those drawings. They offered a solution to the problem of laying pipes over the uneven and shifting bed of the river. Ann was much amazed that I decided to visit the fish market. I assured her I was buying a lobster which she could give to Cook but first I would examine the flexible fan joints of its tail for any guidance Nature herself may offer. Nature proved true, as she often does, and the flexible design was most successful. I am sure if I was to buy a ferret I would find the piece of handsome plate the Waterworks insisted upon giving me! But I will not let Ann search for the drawings.

She cannot come a yard close to the workshop for I fear she may take it upon herself to enter and, seeing what she would call a Calamity, set about cleaning and tidying as if the Good Lord himself were coming to judge my worthiness for Heaven. I have insisted that it is always to be kept as I leave it because nothing will ever be found if the organisation of my mind is disturbed by a system alien to my needs.

The thought of other hands rifling through my papers is a horror to me. Their meaning has grown larger as my mind is plagued with notions in need of fulfilment and I rely upon them more and more. Yet there is a price to pay for keeping papers that are better left folded for the eyes of ages to come. Yesterday, by accident, I came upon Gregory's first papers on Geology which were read at the Royal Society making his cheeks flush with pleasure and his health, for a moment, appear robust and certain. They were with his letters of his travels, letters his brother and I shared, but not opened since. I have hidden away Jessie's little mementoes, such as they were, for they held her small soul and threatened to enfeeble me with grief. That my children were not to last can be like black salt on an old wound, aching when the weather aches, sending sleet into every torn crevice of memory if I would let it. It is a pleasant repast to read Amelia Opei's fancy novels for I cannot pull myself away once I have entered into the first page! But whatever my past times, or the musings of an afternoon, to have fathered seven and outlived all but one is a burden of life no calculation will fix.

*

66

'a transport never before thought of'

George Stephenson
Lime Road Cottage
Killingworth, 1815

I walked this day to Long Benton to clear my head and visit the grave of my dear wife where she lies with our child taken from us as she opened her eyes to the world. I took Robert with me, for he is a strong, strapping lad with a step that keeps me honest. He is home from Percy Street Academy and speaks like a gentleman, much to my liking.

He is full of questions and I am full of answers, for it is a great pleasure to me to have a son at all and a blessedly lively one.

Robert asked of the work hereabouts on locomotives. I happily reported that Tyneside grew in importance since Matthew Murray and William Hedley had arrived. I asked if he remembered me fixing clocks and mending shoes and to mark that, for at Killingworth our future was about far greater things, and would rival Birmingham's one day. I joked that Northumberland had become a veritable college of steam engineers and engine wrights running their eyes over each other's contraptions like so many farmers at a beef sale yard. Robert listened with his head up as I matched his stride.

It is a proper thing, I told him, that men watch their Patents as they have always done but there is a Brotherhood among steam men and we talk well of any engineer's advances.

We made short work of the few miles to St. Bartholomew's, the weather fine and the clouds high and short of rain. I stood very still and said a prayer at the grave. I ran my hands over Fanny's headstone and said that God's weather was keeping it well washed and his mother would like that.

Robert would not touch the stone. He asked if I remembered his mother's cough that troubled her so.

I hope you remember her loving kindness, I replied, and took his arm and turned back to West Moor.

My heart was heavy for I did not know if I had burdened the lad in the cause of loving duty. And it is a truth that I carry, if I did not have Robert, I would willing lie myself down beside Fanny and our precious baby once so light in my arms.

But we had not gone far when he spoke.

I hear Murray has honourably paid Trevithick royalties for the use of his Patent. Indeed, he has, I answered much surprised at his knowledge.

And Trevithick is a genius but a cranky tosspot?

Do I pay good money so you can learn to speak like that, I replied quite nonplussed.

Yes sir, you do, Robert answered quick as a flash, but only if I say it properly. I had to laugh for it was indeed true. How a man sounds is held for him or against him and I will not have Robert burdened with the agricultural vowels of the North.

Do not think poorly of Trevithick, I said. His engines are remarkable for their Audacity. He has one in Cornwall, Wheal Prosper near the sea, and he reported to John Rastrick that his steam poll engine, using expanding steam at 100 psi, has a Smeaton Duty equal to 40 million pounds of water lifted by one foot per one bushel of coal used.

That is a mighty duty, Robert exclaimed.

Yes, it is. Trevithick is a fine engineer but he pushes steam to extremes and I am doubtful his boilers will hold. Hereabouts we go step by step, I cautioned.

But he has adaptations aplenty said Robert, suddenly enamoured of his Cornish pisspot.

How so, I asked curious as to what he would say next he seemed so sure of himself.

He made the return-flue boiler, he replied and quickly added, he put the steam jet through the chimney and connected his wheels.

That is true, I replied, quite amazed. Is there anything else?

Yes father. There is. His engine was too heavy and broke its running rails.

Indeed it did, I replied.

We will have to attend to that, said Robert, almost to himself.

Matthew Murray is our local genius with engines. Perhaps you should visit him, I said.

Do you mean Murray's improvements to William Murdoch's D slide valve, Robert asked.

What is this D slide valve you mention?

I have not seen one, said Robert, but I have been reliably informed it allows steam to enter both ends of the cylinder, by sliding backwards and forwards, which is a good thing.

You have been reliably informed, I replied. Murray arranged his valves to be attached to gears driven by the rotary shaft of the engine. And how might that be a good thing?

Robert laughed. Father you are testing me like one of your workers, he exclaimed.

I have an answer, I said, if you do not have one readily at hand.

Murray aims to use the engine's power to aid its own efficiency by driving the valves, he replied, most adamant.

I laughed and patted him in good humour and told him he sounded like a son of mine to which he replied that was because I sounded like a father of his.

We spoke of many things as we ambled back to Lime Road, for the day was still with us and the sun warmed our faces. I told him how many were in awe of the success at Middletown Railway where Murray was the first to place a piston in a horizontal position and patented an automatic damper that regulated the air to feed the furnace.

I explained that an Engineer always worked to please the Goddess Simplicity and her Sister Efficiency, and Murray had done this by allowing the boiler pressure to control the damper.

Are you then, Father, in great admiration of Murray?

I am in admiration of any man who advances steam power and Murray is certainly one.

He asked about Murray's Salamanca engine. I described how it was working from Middleton Colliery along the old wagonway to Leeds harbour and had pulled 8 wagons of coal.

It was the first rack and pinion locomotive, was it not?

Yes, indeed it was. John Blenkinsop has rightly taken out his patent for his rack propulsion but I believe he is up an alley when we need a high road. His system may serve for steep grades, but would not stand much wear and tear and we will need another way.

I explained to Robert that steam men were in general agreement that our engines can pull about four times their weight by what we call adhesion.

Iron wheels on a dry iron track, Robert replied.

Yes indeed. But Blenkinsop wants his engine, weighing at five tons, to safely haul a weight of ninety tons and so he devised a rack outside the track driven by a large cog wheel, but it will not last.

Quite *pesle mesle*, (1) Robert replied, smiling at his French, adding, too much haste not enough simplicity.

It was the first time I saw him pleased with himself since he arrived home. I would have been content to listen to its echo back to West Moor but Robert,

I found, could talk of locomotives at a pace to equal his father. He enquired more about our competitors.

I explained Murray was full of innovations and has made his piston crossheads slid in guides instead of the parallel motion linkage as is usual practice. But more importantly, Murray's engine works for a profit and presently they have two more engines working. They have replaced thirty horses and have let go over one hundred men.

Robert was silent and I enquired of his thinking.

What shall those men do for their livelihoods?

His question surprised me and it hung in the air unanswered.

My work was with managers and the collieries they ran and I had left such considerations to them.

Mining men are a hardy breed, I replied presently. Our steam machines have enabled us to expand our coal production so we are, more and more, in need of miners.

And how shall they get themselves from their villages to their new work?

Indeed Robert, I thought, how shall they? But a reply fell from my lips, imagined from what I do not know.

They will adapt, I said. Look you at the birds. They flourish without a tree in sight.

That is true father, he said. But I fear the birds are far in advance of us. We have yet to grow wings.

He had me laughing, and at myself.

Son, I said, wiping my eyes, you are like a Meadow Pipit, a tiny chap who spends the whole of summer announcing his arrival and daring the Merlin to make something of it.

Sorry father, I should not ask so many questions.

In a fine humour we walked on, the late afternoon full of shadows and fast giving way to evening. I heard how steam has become 'quite the thing' and many of Robert's friends would walk miles to watch a 'Steam Carriage' puff away but it takes a mighty friendly fireman to cajole folk to climb aboard although he would not be afraid to do so.

If I may ask, Robert said, there is something which I marvel at but cannot explain.

And what is that, I asked wondering what the boy's mind would advance next.

It is a strange truth that when one sees a great design it appears so obvious as to have been always so, he said.

I did not reply with ease for I was uncertain of his meaning. I could only reply that before any advance made in Engineering nothing was obvious and that often it was trial and error that led the way, and often then, the outcome could be a happy Accident.

I knew you would say that Father, he said. You do not know how great you are.

I wished his mother was here so she could kiss him for me, he being so tall and bright, but I was lost in the jumble of my reply. On approach to West Moor silence came upon us, very light and steady and companionable. There was a chill wind blowing by the time we returned and my sister fussed over us and insisted we sat by the fire and drank a good broth. Robert is now abed and I am sat most happily with my thoughts.

It is time to smoke my new clay pipe, a gift from Robert who proudly assured me it was the latest thing for 'tobacco drinking' and was fashionable enough to suit any man, woman or child. I do enjoy the solace of a pipe and contemplated as I tasted its sweet aromas, if I was such an omnibus of a man I suited all categories!

He reminds me of his dear mother, he is so open and honest with the world. I must remind him that although there is a fellowship, nay, a brotherhood among steam men, there is also competition for we all have families to feed and reputations to make. But there is no better business in the world, and besides, it is well too late to turn him from it.

*

It occurs to me that I must have slept, for the wind has died and the cottage has settled most comfortable. Except for the homely snoring of sister Eleanor, there is not a creak or rattle to be heard. Outside is the regular muffled chug of the colliery pumping machine so deep in our bones we hardly hear it. I should be to bed but my thoughts leap about and beg to be put in order. It is often so when I am with Robert. He does not know how he tests my mettle with his unquestioning belief that where there is a problem I will find the solution.

He reminds me of Will Hedley whom he met when he was a little terrier running everywhere and yet to be schooled. Will boasted he had been 'growing boys' for years but they were presently at school. He took Robert by the hand and conversed with him most seriously as if he was already a philosopher engineer like himself. In the years since Will has certainly made a name for himself at Wylam Colliery. His locomotive has replaced horses on the wagonway and pulls the coal chaldrons at 5 miles to the hour to the docks on the Tyne. He has let it be called *Puffing Billy* which Robert curiously tells me is a most happy name.

I will tell anyone who asks that I am in admiration of Hedley. I once stood with him in the rain which he seemed not to notice and heard his ardent praise of his *Puffing Billy* spitting and hissing most happily in front of us.

I see your cylinders are vertical either side of the boiler, I noted, and they look to be about 9 in. by 36 inches.

He raised his eyebrows, about 9 in. by 36 inches, you say? Have you been creeping about with your ruler? He was quizzing me but he looked most pleased and patted his boiler, with gloved hands I must note, as now we all do.

You are most cannily correct, George, he replied. They drive a single crankshaft for stability and better traction. My steam pressure is at 50 psi. and she pulls well. He looked most pleased with himself as I would be.

I think you have done well to leave it to adhesion instead of cogs and racks, I offered, but did not add that Blenkinson's system was the most tortuous idea I had seen in these parts.

Will sighed, adhesion is the way forward, he said, but looked mournfully at me. She is a beauty but she is at 8 tons and has already broken the wagonway plates. He looked most serious for a moment but could not help a laugh as if he was, nevertheless, proud!

If you add more wheels and spreads your weight you should be better situated, I offered.

Exactly my plan, George, if you don't beat me to it. I must warn you, my Patent is on its way! I laughed at that and told him he would be a witless engineer otherwise.

I did not mention that Murray at Middleton had challenged me that my *Blucher* is merely a cousin of his *Willington* working at Coxlodge which I took as a slight. It was a fruitless challenge for I have usurped his family, *Blucher* being the stronger relative, able to haul 30 tons of coal up our hill at West Moor at 4 miles to the hour and pulls coal from Killingworth to the Wallsend coal staithes.(2)

There is not an engineer will make an engine without a name and I was so bold as to name my *Blucher* after the Prussian general celebrated for his timeliness and courage at Waterloo. It is a famous victory and General Blucher's happy arrival at the battle made Wellington a most lucky man. Like Hedley I will have nothing to do with cog and rack pinions, instead I have flanged the wheels and rely on adhesion to hold *Blucher's* wheels to the track. Robert's Uncle James asked to be *Blucher's* first driver, whether to honour the General or the engine I have yet to determine. I must tell Robert that his Aunt Jinnie has been up at dawn to start *Blucher's* fire and James tells me she had occasion to push the engine over a difficult turnpike.

But perhaps not, for Robert would quiz me as to why that needed to be so. He already knows the running of any engine tells a story and to listen for its temper and heed its chatter. *Blucher* suffers, as do all our engines, with being short of steam when it needs power.

Like Will Hedley, my daily meditation is the problem of transmitting power more evenly to the wheels. I aim to make the action of the engine regular and *Blucher's* two crankshafts assist well, but the connecting rods and driving wheels are not evenly in rotation, causing alternating pressure on the cogs and wheels and James reports that they make a jerky and noisy motion. Jerky and noisy! This will undoubtedly get worse as the cogged teeth become worn. I knew this would be so but I let it go for my head was brimmed full of other calculations. It is one thing after another, niggle after niggle, and there is no rest for a steam man.

We have constant problems sealing our pistons. Our engines are too heavy. If it were not for the strong shell of our boilers our engine frames would buckle. Steel to make springs is either too high in price or too low in performance. No matter the success of Hedley, Murray and myself, we are a long way from inducing more collieries to use locomotives instead of horses. The cost of coal will hamper these locomotives. Now corn is cheaper because the war is over, horses will prove their usual worth. I sense that my colleagues will let their engines run as they are which will not be enough to advance steam power. But, notwithstanding all that I write, it is my belief we are at the beginning and not the end of locomotive power.

Whether I build locomotives or not steam will advance and become the greatest traction-power ever made. I see it clearly in Robert. The young are not burdened with the past and take it for granted our engines will pull a weight and size never before imagined.

I am beholden to Nicholas Wood who allows me use of the colliery workshop handy behind our cottage, but Robert sees him as keen for business and has well reckoned *Blucher's* power to Wallsend staithes.

Nevertheless, we are blessed to have a man whose mind is open to notions that others hereabouts consider foolhardy for there are stories of death and explosions many say are Retributions against the madmen who build such monsters. I know men who will say to my face my engine is but a Curiosity, and not even a useful Curiosity, for its only entertainment is the suspense of waiting for a terrible blow-up. There are learned men who think it impossible that a carriage could be drawn along any path at the rate of 12 miles an hour without a horse!

I do not allow such nonsense to stand for such a speed will come to pass. I say to anyone who doubts the progress of steam that they should look back in wonder at our advancements where we have safe and reliable 9 in. cylinders at 50 psi when barely 20 years ago we had 25 in. cylinders at 10 psi. Whatever men may say of Trevithick, his Wheal Prosper engine has long since trebled the working power of the old Boulton & Watt whim wheel engine at Carharrack. He sets a fine pace for us to follow. He has never believed anything other than we will master steam and it will be a fine servant for mankind.

There is another problem which nags at me claiming a great Necessity. Shall we be making Engines only to run around their small worlds, trapped on the wagonways of local collieries? If we are making traveling engines then they should travel! Most collieries are following the old horse ways and hereabouts they are either 4 ft. 4 in. to 5 ft. across. Wylam and Middleton run at 5 in; at Killingworth I am using 4ft 8 in. I am ready to adjust, for if we have any common sense about us to advance locomotives, we must come to agreement on the gauge used. Different gauges in different towns will make of engines merely local toys and an engineer will find he can only run around his back fields. We need a uniform gauge, then we will see what steam can give. I have no doubt of it.

As a useful distraction, I have lately taken up the task of protecting the miners by way of a safety lamp, if I can make one burn in a gaseous atmosphere without blowing their heads off! If he reads this Robert may take me to task but I am not so flippant as my words, for the names of good men lost at Killingworth number over 20 these last years and the faces of their wives tell a story few care to read nor do they care to recall how the earth shook from the deadly explosions at Felling and the horror could be heard miles beyond Killingworth. They counted the missing as 92 men and boys lost to their families.

The night is old, and sleep knocks at my mind most put out it has been denied. I will speak to Robert tomorrow of the talk among steam men which is particular not to the moving of coal, but rather, the moving of people. We aim for an engine pulling comfortable wheeled coaches, free of the vagaries of roads, and able to run in all weather.

I can write the following for it is my clear and steady belief.

If we can make it safe and true we will offer a transport never before thought of and never before enjoyed by working people over such speed and distances barely to be imagined.

*

67

'memories grow like fungus on my wall'

James Watt
Heathfield Hall
Handsworth, 1818

Jim has written with great news. He no longer surprises me with his adventures and daring. He writes from Coblenz where he has moored the *Caledonia* and says he judges his is the first English steam ship to have crossed the Channel and forged up the Rhine. He says it was a joy to feel such power under his feet and even tho' Coblenz is a most ancient city the good Burghers thought him more than 'modern.' He is so far away I fear I will be alone in the world. Ann reminds me I have not wanted for visitors, but of late my preference lies with family or old friends.

In my reverie today, I would have my long-departed friend Erasmus visit. I have much to discuss and I do not imagine such a man would have a woebegone spirit and stink of the grave when he died with so much life in him. I have secreted away some biscuits, which I keep airtight, and a more than half good port and it amuses me to think, that this is all that would Suffice to lubricate our discourse on matters which take our interest. For instance, I would want his opinion how it is that fungus grows and multiplies under certain conditions and if that proliferation could effect a change in the mental process of living things about it. In short, I would ask if it were a possibility that memories grow like fungus on my walls, for it seems to me not a coincidence that the two arrived together.

It is also worth noting, as I have over these few weeks, that the fungus seems to reproduce of its own accord and needs no association with others of its kind to do so. My memories flow with much the same abandon, often having no association with their fellows but being vivid nonetheless. As a man privy to the secrets of various plant species Erasmus would surely enlighten me as to the reasons for this peculiarity. Perhaps not of my memories but certainly of the fungus which has settled on my southern wall.

I remember the afternoon we met his nails were black from his garden and he looked more like a farmer than a learned doctor. He was an imposing man with his large hands and his huge goodwill towards the world and all its marvels. I would discuss with him my wish to be buried at Handsworth Church, where dear Boulton is buried rather than home at Greenock, and I am certain he would nod and pat my arm and tell me to do so, but I should be prepared in case Boulton would be in my ear for all Eternity about some scheme or other!

Erasmus is greatly missed in this workshop, as he is in many others, it being a long decade since his passing so it was both a welcome surprise and a tender moment when his son Robert visited Heathfield Hall with his youngest boy Charles. It was, in my reckoning, several weeks past but most vivid still. Charles shook my hand most properly and immediately told me he was a boy of nine summers and that his mother had gone to Heaven. They were on their way to Shrewsbury School where the older Darwin boy is already boarding and Charles was to be enrolled. Robert is already a remarkable copy of his father, his girth expanding to meet his fine good humour and largesse of spirit, although I noted his paleness and put it down to the recent loss of his dear wife Suzannah whom I had known as a sweet child when she was Suzannah Wedgwood. Charles seemed not of that generous mould, as he is a slight brown-haired button of a lad who wore a frown of concentration as if he knew already what it means to be a motherless boy.

He then surprised me by his enquiry concerning the fungus, asking me why I tolerated its presence and was much taken by my answer that I was waiting to learn about its habit of procreation. He nodded in approval and voiced a confident opinion that his grandfather would have known the answer which caused his father and me to catch each other's eye and laugh at his certainty. Indeed, said his father, you may very well have named the one man who could have solved the mystery.

And since your grandfather is no longer with us, you may have to solve the puzzle for us, I added, being very pleased with the lad.

Oh sir, he replied, I am only just on my way to school and may need a few years before I am that clever, making his father and I laugh again.

It was a most pleasant visit but it left me drifting with the rest of the afternoon, where I could not find a fascination with any one thing about me, and came to dwell on the fungus and my dear friend Erasmus and the questions I would put to him.

***

1830

*Robert Stephenson
Rocket locomotive 1829*

In 1829, the company Stephenson and Son entered the Rainhill locomotive trials conducted by the Liverpool & Manchester Railway. Robert Stephenson's *Rocket* won with speeds reaching 29 mph. So great was the public interest in steam locomotives that over 10,000 spectators cheered on the engines. Passengers experienced speeds entirely new to humanity where their eyes confronted the world as one long continuous line, and familiar objects blurred beyond immediate recognition and their heads whirled and they thought their sight under threat. The *Rocket* was taken as the model upon which future steam locomotives were based for more than a century, and that seems a reasonable time to say steam power had come of age.

The astonishing human project of steam power, nurtured in the rigor of *Nullius in verba*, grew from a rudimentary vacuum pump to an engine capable of moving hundreds of people and tons of goods over long distances. It could out run and out pull any horse known to man, mile after mile. It was built on dreams of advancement not equalled until the Wright brothers took to the air in 1903. This perfect storm of engineering power has a remarkable geography for, as the following will tell, the bones of its pioneers lie within a 150 mile radius north and south of Birmingham, a proud market town whose merchants had made it prosper.

It is not always a story of happy triumph. In 1780 Matthew Wasbrough designed and built the first rotary steam engine using the efficient and simple crank and fly-wheel. A year later he was dead at 28, driven into illness by exhaustion and despair. He is buried in Bristol. Richard Trevithick died of pneumonia, penniless and alone in 1833. His fellow workers at Halls Works paid for his funeral and a night watchman to guard against body snatchers but, being a contrary visionary, Trevithick may have willingly given his body to medicine. There has been no way of tracing his unmarked grave at St Edmund's Dartford, but in 1888 a stained-glass window in his memory was presented to Westminister Abbey.

John Smeaton is honoured for his extraordinary engineering prowess in building lighthouses, canals, bridges and harbours, as well as his work in steam. He died aged 68 in 1792 and lies in the parish church at Whitkirk, Yorkshire. In 1994, in commemoration of his achievements, a memorial stone of Purbeck marble was unveiled in Westminister Abbey. It includes a bronze inlay of Eddystone Lighthouse, a revolutionary restoration that made Smeaton famous in his time.

Matthew Murray, who made the first commercially viable steam train and became a successful and respected engineer, died aged 60 in 1826. He was buried in St. Matthew's churchyard, Holbeck and lies under an iron obelisk made at his Round Foundry, where his *Salamanca* was made. The Round Foundry has been both preserved and absorbed into modernity as a Media Centre hosting various creative and digital media companies. You can visit history and have a coffee at the same time. William Hedley of *Puffing Billy* fame, lived to see the great success of the steam locomotives and the phenomenal expansion of the railways. He died in 1843 at Burnhopeside Hall, and was buried at Newburn parish church.

Matthew Boulton and James Watt became celebrated men and died honoured by their nation. They produced 500 engines altogether, from 1776 to 1800 and Watt's innovations and inventions transformed a rudimentary steam pump to a versatile transportable prime mover. By today's standards they were millionaires. You may have seen their stern and slightly harassed features on the £50 note. Watt survived Boulton by 10 years dying in 1819 at the age of 83. His son James Jr. had a mistress but never married and devoted his life to the company. He died in 1848, but James Watt & Sons continued to make engines until 1895, with cylinders over two metres in diameter cast from 33 tons of iron, smooth and airtight, with a piston pulling over 125 horsepower.

Having spent his creative life in the shadow of Boulton & Watt, William Murdoch now stands their equal in gilded bronze larger than life in Birmingham's Centenary Square. Murdoch died in 1839, aged 85. He left a house with advanced innovations including gas lighting, a doorbell worked by compressed air and an air conditioning system. He chose to be buried in the same ground as Boulton and Watt, *St. Mary's Church, Handsworth*.

George Stephenson died in 1848, disillusioned, as was Denis Papin many years earlier, by the aristocratic establishment which controlled the Royal Society and where it was simply not possible that a foreigner or an uneducated lower-class northerner whose parents were illiterate could invent a safety lamp as good as, if not better than, a recognised scientist. Despite the support of friends, in Stephenson's case it was suggested that he may have copied the idea of a safety lamp from Humphrey Davy. Stephenson was buried at Holy Trinity Church Chesterfield and would have been surprised to learn the Bank of England had placed his image on the £5 note. Robert Stephenson outlived his father by only 11 years dying in 1859 at 55 from a kidney illness brought on by the working conditions he had suffered for decades. His nation mourned him with a grand funeral and took him to be buried in Westminster Abbey.

*

EPILOGUE

So here we are, so short and so long a time from the enterprising optimism of the 18th century where scholars pondered Adam Smith's *The Wealth of Nations*, where prosperity for all was possible. Today we reluctantly read Tony McMichael's *Climate change and the health of nations: famines, fevers and the fate of populations*. It is a great historical irony that Watt's steam engine, by its savings in coal fuel, lead to a revolution based on that ubiquitous fossil fuel. But it did and here we are.

Without so much as a by our leave, science responds to the pressing necessities of its time, exploring and creating wherever its resources permit. It sits apart from the moral implications and unforeseen consequences, leaving such considerations to us. Like an addict who takes their first drink, who were we to know that fossil fuels would be our poison. And how are we to break our addiction by reason alone, when pain beyond endurance is the only known breaker of addiction? Ironically, the power of science that solved one painful necessity tells us another is on its way.

It is strangely moving to learn that a month before James Watt died, a baby girl was born on the east coast of the newly united states of America. Thirty-seven years later, when the railways were driving industrial expansion, this grown woman gave humanity its first warning of the dangers of carbon-dioxide. Eunice Foote (1819-1888) had a paper read at the 1856 American Association for the Advancement of Science (AAAS) where she correctly hypothesised that carbon-dioxide gasses could trap heat in the Earth's atmosphere. By 1859 John Tyndall had published a series of studies verifying this hypothesis. One hundred and fifty-nine years later we still struggle with the implications of such knowledge as if science had never quite proved itself and we orbit on some other planet, free of consequences.

Despite this, like a continuous breaking dawn, science moves on. Barely sixty-four years after the death of James Watt, the Renewable Energy team made their first mark. Charles Fritts, (1850-1903), created the first working selenium cell in 1883, and only five years later Charles F. Brush, (1849-1929), built a 12kw wind generated turbine. The modern Danish V164 wind turbine, with 80 metre blades, produces 260,000 kw over 24 hours.

In 1941 Russell Ohl, (1898-1987), patented the modern silicon solar cell. In the early 1980's John B. Goodenough pioneered the lithium ion-battery and battery research presses forward, insistent and optimistic, year by year. The efficiency of solar cell technology has been advanced in 2006, 2008, 2012 and 2016. Concentrated solar power systems (CSP) are experimenting with a variety of ways of driving modern steam turbines.

In the 1990s, General Motors produced the first viable electric car. Drivers loved it but GM doubted its profit potential. Today they are in no doubt. In 2017 Volvo announced it will 'go electric' and no longer make cars powered solely by the internal combustion engine. Most car manufacturers are developing their alternative fleets of EVs. Elon Musk from Tesla is not the only visionary.

I like to think that James Watt and his colleagues, surprised, but alert to the evidence of science, would have been proud to know their university was 'up with the times' when in October 2014, Glasgow University was the first academic institution in Europe to divest itself from the fossil fuel industry.

Governments worldwide, however, seem caught in a medieval mind split where the rational and the irrational are allowed equal weighting. The modern world easily honours medical science but ignores environmental science. Faced with our disjointed madness, and caught in the comfort of denial we cannot afford to place our faith in old certainties already washing away.

As Erasmus Darwin would remind us, our scientists are strands in a spider's web, seemingly fragile in the morning's sun, but resilient and perfectly formed to catch whatever ideas fly through their minds. It remains for us to listen for science is both the cause and the cure of the dangers we face. The past and present flow together in a long continuum to shape our future, joined by choices we have or haven't made. My story of steam traces 150 years of engineering perseverance from Papin's hissing *Steam Digester* of 1679 through to the 1829 Rainhill Trials where Robert Stephenson's *Rocket* made real to thousands the promise of locomotives. On my calendar 150 years of Renewables gives us to 2033. There may still be time.

*****

SOURCES AND NOTES

3 Robert Boyle London, 1663

p 5 (1) *virtuosi* Latin: very learned persons

p 7 (2) In 1649, only 14 years earlier, Parliament put Charles I on trial, sentenced and executed him.

p 9 (3) 'mufflements' Lancashire dialect for woollen garments

p 9 (4)_ 'rained pikels with the tines downwards' Lancashire dialect: sharp rain like the points of a pitch fork

p 9 (5) In 1677 Towneley began the first systematic measurement of rainfall in the British Isles. He published 15 years of records in the Philosophical Transactions of the Royal Society in 1694. In 1977, British meteorologists celebrated 300 years of record keeping begun by Towneley.

p 9 (6) In 1612 the whole of England was agog at the trials and hanging of 11 witches from Pendle Hill. Thomas Potts wrote the official report. *The Wonderful Discoverie of Witches in the Countie of Lancaster 1613*

5 Robert Boyle 1679

p 14 (1) Boyle's actual list. Royal Society Archives RB/1/8/30 (AFXSA2JSHNHD)

6 Christiaan Huygens The Hague 1690

p 18 (1) *capti per antrum* Latin: trapped in a cave

p 18 (2) Otto Von Guericke (1602-86) *Experimenta Nova* Published in Latin 1672 Magdeburg University. A book of prime importance in electronics, air pressure and the vacuum pump.

p 19 (3) Isaac Newton *Philosophiae Naturalis Principia Mathematica* 'The Mathematical Principals of Natural Philosophy' Published in Latin 1687. Considered to be one of the most important scientific books ever published. Newton explained the laws of motion and gravity.

p 19 (4) *laboratorium* Latin: laboratory

7 Robert Boyle London 1690

p 21 (1) "danger & difficylties" *The Christian Virtuoso* published in English 1690.

p 23 (2) I will not be the first man to note that there is a distance between a Truth that is Glimpsed and a Truth that is Demonstrated.'. French scientist Alexis Claude Clairaut observed, *"what a distance there is between a truth that is glimpsed and a truth that is demonstrated."* (1759) regarding the rivalry between Robert Hooke and Isaac Newton over the discovery of the inverse square law of gravitation. He sided with Newton because his maths demonstrated the truth of his discovery.

p 23 (3) Phillipe de Mornay (1549-1623) French Protestant theologian.

8 Thomas Newcomen Dartmouth 1695

p 24 (1) Red Letter Day: a special day. In medieval times a holy day (holiday) was written on the calendar in red ink.

p 26 (2) John Flavel (1630–1691) Baptist theologian, preacher and tutor. The Baptist were, from time to time, harassed and oppressed.

9 Denis Papin, Kessel 1898

p 27 (1) tragique French for tragedy

p 28 (2) id est Latin: that is i.e.

p 29 (3) *pique-nique* French picnic - a newly fashionable pastime of eating outdoors.

10 Thomas Newcomen Dartmouth, 1699

p 31 (1) Thomas Savery 2 July 1698 Letters Patent 'Raising Water by Means of Fire.'

12 Thomas Newcomen Dartmouth 1705

p 37 (1) Dr. Robert Hooke (1633-1703) Curator of experiments RSL 1665-1703 Philosopher, inventor, clock-maker, architect, polymath. *Micrographia* published 1665 in English. Small drawings of organisms under a microscope where Hooke used the descriptor cell as a biological term.

13 Denis Papin The Dover Road 1707

p 40 (1) La Manche French: The English Channel

p 40 (2) Munden in Lower Saxony was the scene of perhaps the first reaction against steam power. Papin was set upon for daring to 'ferry' the river without permission of the Ferrymen Guild.

15 Denis Papin London 1708

p 46 (1) The Great Fire of 1666 ravaged central London over four days, destroying some 70,000 buildings, homes, factories warehouses and churches. There were only six deaths recorded which seems unlikely with temperatures reaching 1250 O C.

p 49 (2) The Boyle lectures - Robert Boyle left money in his will for public lectures to be held in defence of Christianity.

18 Denis Papin London 1712

p 58 (1) The Old Swan. A public House by the Thames stairs frequented by Watermen and their passengers. Samuel Pepys Diaries1665 mentions making landfall at *Dukes Shore* before wading up the beach to Narrow Street. Thames stairs were often built near a public house.

P 58 (2) *loqui linguis* Latin: speak many languages

19 Thomas Newcomen Birmingham 1712

p 63 (1) Snifting Clack To solve the problem of the cylinder being 'wind-logged' a release valve was placed near the bottom of the cylinder which open briefly when steam entered and non-condensable gas exited the cylinder.

20 Thomas Newcomen London 1719

p 67 (1) Luke 12 *"And he said to them, "Take care, and be on your guard against all covetousness, for one's life does not consist in the abundance of his possessions."*
Newcomen's sentiments on this biblical text expressed in a letter to his wife London, December 30th, 1727 purchased by the Corporation of Dartmouth, March, 1952.

23 James Watt Greenock 1756

p 81 (1) Sir Alexander MacFarlane Glasgow University MA 1728. FRS. Jamaican merchant and mathematician. Donated his astronomical instruments to his alma mater. The MacFarlane family motto was 'The Lord my light; the stars my camp.'

25 James Watt Glasgow University 1760

p 86 (1) *Universal Magazine* known as *The Universal Magazine of Knowledge and Pleasure* - a monthly magazine (1747–1814) Editor John Hilton.

p 88 (2) Jacob Leupold *Theatrum Machinarum Generale* 1724. Leupold's General Theories of Mechanics provided the first systematic analysis of the general principals of machines and was highly influential. It included a high-pressure non-condensing steam engine design, 100 years ahead of its time.

p 88 (3) Jacob Leupold *'Krafft, ohre Kunst, ist hier umsunst'* - strength without art is to no purpose.

26 James Watt Greenock 1762

p 91 (1) poured our milk onto the ground, butter and onions also thrown out. The death spirits were considered very powerful. The rejection of food such as milk and butter were to ensure that death did not infect them and remove the '*toradh*' (fortune) from them. Handfuls of soil symbolised the earth to which all bodies return and the salt represented the eternal soul. (Cailleach's Herbarium)

27 John Robison Glasgow University 1763

p 93 (1) William Harrison, son of John Harrison (1693 - 1776) the master clockmaker who preserved for over 20 years to build an accurate seagoing clock to record longitude as required by the competition set up by the Board of Longitude formed in 1714.

28 James Watt Glasgow University 1765

p 97 (1) Recounted by James Watt to brothers Robert and John Hart. Robert thought Watt 'the greatest and most useful man that ever lived' Robert Hart Reminiscences of James Watt. 1857 *Transactions of Glasgow Archaeological Society* Vol 1.No. 1 p 1-7 1859

29 Erasmus Darwin Lichfield 1767

p 104 (1) James Brindley (1716 -1772) With little formal education but exceptional natural ability Brindley became the master canal builder of his time.

31 Erasmus Darwin Lichfield 1768

p 109 (1) John Mitchell (1724 - 1793) Cambridge educated natural philosopher and clergyman.

34 James Watt Kinneil House 1769

p 120 (1) Richard Edgeworth (1744-1817) Oxford graduate, natural philosopher and inventor, especially in mechanics.

36 John Roebuck Bowness 1772

p 126 (1) the Ayr bank. In June of 1772, Douglas, Heron and Company, known as the Ayr Bank, failed with liabilities exceeding1,000,000 Pounds Sterling.

Founded only three years before with the backing of most of that country's elite, this calamity cut short a burgeoning economic boom short.

p 126 (2) Alexander Fordyce (died 1789) 'One rascally and extravagant banker has brought Britannia, Queen of the Indies, to the precipice of bankruptcy! It is very true, and Fordyce is the name of the caitiff. He has broken half the bankers.' (Horace Walpole, 22 June 1772)

42 John Roebuck Kinneil House 1776

p 149 (1)The Scots Magazine. First published 1739 to offer articles of Scottish interest. Still publishing under that name today.

p 150 (2) *The Improvers*. Founded in 1723 by aristocrats to improve the yield of their lands by the 1760's it was led by knowledgeable middle-class owners organised into societies, spreading information and making great gains for Scottish agriculture.

43 James Watt Brimingham 1777

p 153 (1) Matthew Bolton letter to Thomas Ennis 1776.

p 154 (2) fork Cornish word for lift or pull. In this case pump water out of a mine

p 155 (3) tributers As individual contractors tributers were paid only on the amount of ore they recovered.

44 Erasmus Darwin Lichfield 1779

p 162 (1) Dr Withering (1741-1799) FRS trained Edinburgh University. William Withering was an English botanist, geologist, chemist, doctor and the discoverer of digitalis. Appointed physician to Birmingham General Hospital at the suggestion of fellow Lunar Society member, Erasmus Darwin.

45 Andrew Cosgrove Handsworth 1779

p 171 (1) Surveying the lie of this land. Cosgroves' assessment was correct. In the 1790's work began to lower the Smethwick Summit and remove the necessity for 6 locks, 3 on either side of the summit.

50 Andrew Cosgrove Handsworth 1780

p 191 (1) Birkacre Mill 1779 riots. 'they went in at 2 o'clock and before 4 destroy'd all the machinery, The Great Wheel, and set fire to the broken fames' Home Office papers October1799

51 John Smeaton Ramsgate 1781

p 198 (1) *Conservation of Energy v's Conservation of Momentum*. This controversy raged between the theoretical philosophers and the practical philosophers. Both Newton and Leibniz would be proved correct, but in different ways.

55 Erasmus Darwin Derby 1794

p 216 (1) The Loves of the Plants published 1789. A poem with philosophical notes.

p 216 (2) Carl Linnaeus (1707 - 17778) Systema *Naturae* first published in Latin: Leiden 1735. Swedish botanist who formulated influential classifications.

p 216 (3) Zoonomia published 1794. A poem with philosophical notes.

p 217 (4) Matthew Boulton Catalogue of facts. Matthew Boulton to Erasmus Darwin 4 January 1790 EDL 198 n.i.

p 219 (5) *E conchis omnia* Latin: Everything from shells

56 James Watt London 1795

p 221 (1) *a la mode* French - meaning in the latest fashion

57 Richard Trevithick Camborne 1795

p 228 (1) *tout suite* French, right away, with all haste.

p 230 (2) David Giddy later known as David Gilbert. (1767 - 1839) was a Cornish engineer whose command of mathematics supported Trevithick and others in realising their designs. He became a Fellow of the Royal Society in 1791 and served as its President from 1827 to 1830. He changed his name to Gilbert in 1817.

64 George Stephenson Killingworth 1812

p 256 (1) staithes specially made piers for loading coal onto ships for export.

65 James Watt Handsworth 1816

p 260 (1) Robert Chambers *The Book of Days*. Published 1864 Watt's trip is quoted as an excursion in 1816.

66 George Stephenson Killingworth 1815

p 269 (1) *pesle mesle* Old French: pell mell, a disorderly or reckless hurry.

Original Documents

p 30 Thomas Savery 2 July 1698 Letters Patent 'Raising Water by Means of Fire.'

p 65 Advertisement *'London Gazette'* for 11–14 Aug. 1716 Newcomen and Calley's company, *Proprietors of the Invention for Raising Water by Fire,*

p 114 1769 No. 913 James Watt Letters Patent

p 148 *Ari's Birmingham Gazette* 11 March 1776 Boulton & Watt's first success.

p 164 *Ari's Birmingham Gazette* 20 April 1778.

p 203 *Felix Farleys Bristol Journal* 1781 Obituary for Matthew Wasbrough.

p 204 1782 Specifications of Patent James Watt

p 231 *Ari's Birmingham Gazette* 1 February 1796 Opening of the Soho Manufactory

Their Own Words

You cannot read wonderful words without them creeping into your writing.

p 23 Robert Boyle 'To observe the Universe. as it were a great piece of clock-work'
A *Free Enquiry into the Vulgarly Receiv'd Notion of Nature* London 1686

p 58 Denis Papin 'I am in a sad case.'
Letter 23 January 1712 Royal Society Archives Vol 7, 74.

p 100 Erasmus Darwin 'that all warm-blooded animals have arisen from one living filament'

'Would it be too bold to imagine, that in the great length of time, since the earth began to exist, perhaps millions of ages before the commencement of the history of mankind, would it be too bold to imagine, that all warm-blooded animals have arisen from one living filament, which THE GREAT FIRST CAUSE endued with animality, with the power of acquiring new parts, attended with new propensities, directed by irritations, sensations, volitions, and associations; and thus possessing the faculty of continuing to improve by its own inherent activity, and of delivering down those improvements by generation to its posterity, world without end.'

Erasmus Darwin *Zoonomia* Gutenberg Project XXIX 4.8.

p 120 James Watt 'Of all things in life there is nothing more foolish than inventing'
James Watt writing to William Small 28 April 1769 MBP 348/8

p 126 James Watt 'value the engine as less than a pouch of farthings.'
James Watt writing to William Small 25 July 1773 MBP* 348/38

p 131 James Watt 'as wet as water could make me.'
James Watt writing to William Small September 1773 MBP 348/38

p 160 James Watt 'to the ignorant, who seem to be no more taken with modest merit in an engine than in a man' James Watt writing to Matthew Boulton 1779

p 218 James Watt 'inconvenient for a person who wishes to get at my purse that I should keep my breeches pocket tight buttoned.'
James Watt writing to Matthew Boulton 31 October 1780 MBP 349/22

*Matthew Boulton Papers Birmingham City Archives

ACKNOWLEDGEMENTS

My visit to Birmingham's Thinktank in 2016 was a happy event. I fell in love. It might have been simpler if it was a human but no, I fell in love with the oldest working steam engine in the world. My friends gave up asking why. At my local bakery, I found a surprising and hidden group of men who loved steam engines. To Peter Oakley, Denis Dowell and Jim Durham my thanks for their knowledge, persistence and encouragement.

I returned to Birmingham and explored the Black Country and Greenock and Glasgow and its Green. My heartfelt thanks to the generosity of the Birmingham Thinktank Engine Department who gave me their time and expertise simply because I was 'writing a book'; to the Birmingham Library Archives staff so proud of their treasures; to Patrick Turbitt, a true Brummie, who showed me the canals down secret stairways and took me to the Iron Bridge and the Severn River and Smethwick; to the Turbitts of Holly Lane for their warm hospitality and interest in my 'project'; to Carolyn Mapes for the wonderful day at Litchfield exploring the great city of 'philosophers'; to John Hopkins from the Smethwick Heritage Centre who, having summed me up in a few seconds, offered to show me the site of Watt's engine, the canal, the site of Boulton's factory and where Watt stayed so he could walk to work. We stood at the dusty corner of Bridge and Rolfe Streets and spoke of history alive in our imaginations. In 2017, I was lucky to attend the inaugural Garsdale Writing Retreat where I started the first chapters and was played Debussy's Clare de Lune as a by the by. Magic.

There are many more thankyous, but anonymous, especially to the many people I met in Scotland and The North who showed great interest in my research and spoke of the Industrial Revolution as if it was yesterday via the stories of hardship and courage passed down from great-great-grandmothers. To Pamela Cook and Helen Baggott, my early editors, for their encouragement. And finally, throughout the last two years, my thanks to my dear sister for the constancy of her faith and the toughness of her editing.

*

Judith Brooks

Judith Brooks was born in Melbourne in 1945. Educated at Monash and Melbourne Universities she became an English and History teacher. At 70 she decided to start a literary career on the 'better late than never' principle. She has published collected poems and a book of short stories, *The Dancing Lady*. *Raising Water by Fire* is her first novel. She lives with her partner and two Burmese cats in Barwon Heads on the edge of windy Bass Strait.

www.ingramcontent.com/pod-product-compliance
Lightning Source LLC
Chambersburg PA
CBHW051646040426
42446CB00009B/997